MW00677618

TechnoLeverage

TechnoLeverage

Using the Power of Technology to Outperform the Competition

F. MICHAEL HRUBY

AMACOM

American Management Association

New York • Atlanta • Boston • Chicago • Kansas City • San Francisco • Washington, D.C.
Brussels • Mexico City • Tokyo • Toronto

This book is available at a special
discount when ordered in bulk quantities.
For information, contact Special Sales Department, AMACOM,
an imprint of AMA Publications,
a division of American Management Association
1601 Broadway, New York, NY 10019.

Library of Congress Cataloging-in-Publication Data

Hruby, F. Michael.
TechnoLeverage : using the power of technology to
outperform the competition / F. Michael Hruby.
p. cm.
Includes index.
ISBN 0-8144-0415-4
1. Technological innovations—Management.
2. Strategic planning. 3. Corporate planning.
I. Title.
HD45.H78 1998
658.5'14—dc21

98-35078
CIP

Printing number

10 9 8 7 6 5 4 3 2 1

To my parents,
Pollee and Frank Hruby,
two of the best
teachers and advisors
a consultant could have

CONTENTS

ACKNOWLEDGMENTS

Corporate strategy is a tapestry, woven from many threads and creating a pattern that shows what works best. The individual threads of the corporate strategy I call TechnoLeverage have been drawn from diverse sources. I am deeply grateful to those individuals and groups who have laboriously spun those threads. This is my chance to thank them for their contributions to this work. In general chronological order, here are the threads I can identify that I have used.

During World War II, the field of operations research was developed by mathematicians and scientists as a way to break down, study, and solve complex problems. The problem-solving approaches of operations research, whose generic frameworks I studied much later in graduate school at the University of New Hampshire, provided many routes through the analyses underlying this book. I am grateful to all those who developed this discipline.

Also during the war, the officers and men of the destroyer-minesweeper USS *Hogan*, DMS 6, patrolled two oceans. My dad's shipmates taught him many valuable lessons in teamwork, camaraderie, professionalism, attention to duty, anticipation of difficulties, concentration under pressure, juggling multiple priorities, and (in port) the benefits of song. He in turn generously taught these lessons to me. They have been exceptionally useful in dealing with the problems of workgroups and professionalism in business and technology. I warmly thank the *Hogan*'s crew and the U.S. Navy for the skills passed on to me.

Both of my parents approached the challenges of bringing up and educating their eldest son with an energy and enthusiasm that amazes me even today. More than anything, they have taught me that no obstacle remains for long in front of him who seeks to surmount it. Their resourcefulness in using to the fullest all the tools around them to solve problems has shaped my approach to using everything that a technological capability offers. For their love, friendship, and inspiration, I am forever grateful.

From the many teachers at University School in Shaker Heights, Ohio, where I was lucky enough to study for many years, I learned the values of hard work and scholarship. Between the lines, they taught their students that one can learn any subject by studying it carefully and faithfully. That lesson has proved true in the difficult challenge of managing technology for business purposes. Special thanks to Messrs. Baker, Cruikshank, DeVere, DiBiasio, Hoets, Ingersoll, Jones, Lee, McCrea, Rickard, Riel, Sanders, and Schwab for their efforts on my behalf.

Banking expert E. E. Furash was one of the first to understand in the late 1980s that technology was the most important competitive weapon banks had to distinguish themselves in a world of commoditized services. Thanks to Ed for his stimulating article, which got me thinking about how broad the business impact of technology really can be.

My clients over the years have helped me know them, their companies, their challenges, and their opportunities. In thousands of meetings, hundreds of plant tours, and scores of projects, they showed me the inner workings of technology and business. They have shared their thoughts and hopes with me to an extent that still surprises me. I am deeply grateful for the trust they have placed in me and my firm and for the patience they have displayed with what must be our many shortcomings. Special thanks to Dick Detrick, opportunity finder extraordinaire; longtime client Charlie Miller; and Dan Teich, who never tired of asking me when "my book" would come out, even when there was no book.

My family has given me their support, confidence, and love without pause. For my wife, Leslie, and her fascination with technology and management, thanks for her original insight that led to concentrating in this area over all others, and for talking strategy while on our morning runs. For my daughters, Emily and Pollee, thanks for their cheery patience with this project and for the companionship of many evenings as we studied and wrote together in front of the fireplace.

For twenty years, I have read *Value Line* religiously. Their reports are the basis for many of the financial observations in this book. I appreciate their stewardship of "the numbers" on thirty-five hundred publicly traded companies.

My colleagues at Technology Marketing Group have given

me years of stimulating discussions of technology, management, strategies, tactics, and the innumerable paths forward for clients and TMG alike. Thanks to all of them for doing so much of the work, digging up the facts that our consulting work is based on. Todd Himstead invariably comes up with a new twist that solves tricky problems, while Mark Lutz is always willing to offer suggestions for improvement; both contributions are acknowledged with thanks.

To my agent, Mike Snell, thanks for his diplomatic observations about the book ideas that weren't very good, and his laser-sharp perception of one idea that was. For Tom Gorman, huge thanks for helping me organize, balance, produce, and edit the ideas and experience of twenty years of consulting into something manageable, and hopefully informative and interesting. Without Tom, this book would still be piles of little notes and sketches. And to my two editors at AMACOM, Tony Vlamis and Ray O'Connell, thanks for helping me first to get started and then to get finished. I am grateful for their help and flexibility.

Lastly, thanks to the rest of my family and friends for their perpetual interest and respect for my endeavors. Their warmth and affection allowed me to add this book on top of an already full schedule, and they helped me believe at all times that it could be completed.

CHAPTER 1

The Profit Machine

Understanding Technology-Based Strategy

This book is about how to make money with technology. The surest path to that goal follows a corporate strategy I call TechnoLeverage.

TechnoLeverage applies technology to a company's business to lift the financial performance of that business to new heights and outdistance the competition. It helps companies use technology more effectively than they do now to improve corporate growth, profitability, and value. A few companies use this strategy now. Many more can use it if they know what it is and how to go about it.

How sneakers were turned into athletic shoes is a clear example of applying technology to a business to lift its financial performance. Let's look at the product before and after the application of technology.

First, pretend that you are living in the 1960s and that it is a typical Saturday afternoon. Look down at your feet. What do you see?

If you're a man, you probably see a pair of black or white high-top canvas sneakers. They have round ankle patches and square heel labels that say Keds or PF Flyers or Converse. A simple stripe borders the sole, which features a shallow ribbed tread. If you're a woman, you probably see a pair of white or navy blue tennis shoes that lace up or else slip on and display the trademark discreetly on the heel. Your sneakers have two decorative ridges bordering the sole, which has a corrugated tread.

These sneakers and tennis shoes come in one fabric, two designs, and three colors. They are probably made in New England and their prices are low compared with those of other footwear sold in that period.

Now let's fast-forward to this Saturday afternoon. What do you see on your feet?

You're probably wearing running shoes or perhaps court shoes or walking shoes or basketball shoes or cross-trainers. They could be low-cut or high-topped, but either style is lighter than the old sneakers were. They come in any color you can imagine and some that you can't. They carry large logos for such companies as Nike, Reebok, or New Balance; their sides display all the swooping creativity the caffeinated mind of a graphic artist can muster.

These shoes not only look different, they feel different. Sneakers went on your feet and covered them. Athletic shoes cradle your feet. Some have systems of inflatable air bladders or gel in the soles to give you power and protection. The tread on the sole reflects the purpose of the shoe. This is *equipment*, not clothing.

We're not in canvas anymore. (Sorry.) Athletic shoes are made of cotton, nylon, leather, vinyl, or some other material on a rubber, nylon, or composite sole. These shoes probably were not made in the United States and may cost more than the other kinds of shoes you buy.

Athletic shoes have evolved from the humble sneaker into a cultural phenomenon. They've even replaced high heels for women walking to work. They are also a business phenomenon, racking up billions in annual sales through specialty retail chains. Although a tremendous marketing effort has gone into athletic shoes, so has a tremendous technological effort. That technological effort has improved the product itself. In fact, the technological improvement preceded the marketing effort.

Wait a minute: technological improvement? We're talking about shoe companies, not pharmaceutical, microprocessor, or telecommunication businesses! What technologies could have played a role in transforming the sneaker into the athletic shoe?

New Uses for Technologies

Makers of athletic shoes have drawn from many technologies: adhesives, biometrics and foot morphology, ergonomics, design, dye and pigment formulation, foamed plastics, injection molding, insert molding, materials compounding, and synthetic fibers. These esoteric technologies complement the more traditional manufacturing disciplines of industrial engineering, materials planning, assembly, packaging, quality control, shipping, and order tracking.

In creating distinctive, much-improved shoes for diverse purposes, athletic shoe companies applied their own and others' technologies to their products. Wielding all the technologies they could, they improved appearance, utility, comfort, and performance to completely transform and redefine their product. Then they used marketing techniques to create and tap new markets for the products that technology helped create.

It may already be clear that I define technology broadly. My working definition of technology is *the application of scientific and engineering knowledge to solving practical problems*. This definition encompasses a broad range of human accomplishment. Of course, it extends far beyond just information technology, which is what many people think of today when they hear the word *technology*. Although some examples in this book are drawn from information technology, more are drawn from the wide variety of technologies mankind has developed.

The athletic shoe industry illustrates what I call TechnoLeverage. This industry managed to wedge technology underneath its business, to lift it to new, significantly higher levels of size, visibility, profit, and value. Lift is the essence of TechnoLeverage. This book is about developing and executing sound corporate strategies that use technology as a managerial tool. I call these *technology-based strategies*.

In my experience, fewer than 5 percent of all companies, including many that make and sell technology, use a real technology-based strategy. Instead, most rely upon marketing-based strategies geared to winning the battle for market share, or operational strategies geared to cutting costs, or some combination of the two.

Marketing and operational strategies certainly have their place, but they flourished in a different economic environment. Although marketing-based strategy served practitioners well in quieter times, as the experience of most former sneaker manufacturers shows, it can leave a company vulnerable to technologically sophisticated competitors. As for the operational strategy of cost cutting—which hardly amounts to a growth strategy—its limits have become widely apparent.

Is it possible that a third approach—technology-based strategy—could produce superior results? Is there a model for companies to follow to focus primarily on achieving growth by adding extraordinary value? How can a business systematically create value? Most important, after three-quarters of a century of marketing-driven strategy and more than a decade of intensive reorganization, downsizing, and similar operational tactics, what should management be doing now to earn more money and build more wealth for shareholders?

This book addresses these questions. It presents an approach to strategy and management that differs from classical marketing and operational approaches. Marketing and operations are business processes that generate the greatest long-term results when they are applied to something valuable. Technology provides the improved value from which marketing and operations blossom.

Technology-based strategy is readily combined with marketing and operational skills. Such strategic combinations have been achieved by Nike, which has wedded technology strategy and marketing strategy, and by Intel, which has joined technology strategy with both marketing and operational strategies. Technology-based strategies have been carried out successfully by companies in many industries, as we shall see.

In today's rapidly changing environment, it is technology more than marketing and operations that enables a company to become and remain a profit machine. Corporate strategy must explicitly address the issue of technology.

Why Technology-Based Strategy?

This book has three premises:

1. Technology drives business opportunities today.

2. Few companies fully exploit the value that technology can bring to their products and services.

3 Companies need a strategy that helps them systematically create and capture value.

Although these premises seem to be factual and straightforward, they are not widely recognized. Yet they apply to all types of businesses. Whether inventing new technologies, manufacturing technological goods, selling technological services, integrating the technology of others, distributing technology, or just plain using it, every company can benefit from technology-based strategy. Therefore this foundation for TechnoLeverage warrants elaboration.

Technology Drives Opportunities

It may be so obvious that technology drives business opportunities today that many readers may believe they are already operating on this principle. After all, most of us have benefited from the exponential increase in the power of microprocessors and advances in other technologies. Perhaps you work in an industry that microprocessors have spawned within the larger field of information technology, itself only one of many technologies. Long before the first punch card was processed, technologies such as the wheel and the loom, the cotton gin and the steam engine, the telephone and the printing press were creating vast business opportunities.

Fully understanding that technology drives business today means recognizing that because technologies are so numerous, pervasive, and malleable, they enable managers to improve *any* business. Technology drives business first in the sense that a new technology such as aviation or biotechnology can create a new industry. Secondly, technology drives business in the sense that some technologies among the huge variety now available can create new opportunities for any business. Thus any black-or-white distinction between high-tech and low-tech businesses is now artificial and misleading. That false distinction causes managers to miss significant growth opportunities by thinking, as many do, "We're not high-tech."

The outdoor advertising industry, which still consists mostly of billboard operators, clearly illustrates the broad potential of TechnoLeverage. People think of this business as low-tech, based as it is on plywood, paper, paint, and lighting. By late 1991, the outdoor advertising industry appeared to be dying. With as many as 30 percent of all billboards empty of paying messages, excess capacity weighed it down. The costs of putting up advertisers' messages were rising, and ad rates were falling. Billboards depended on the "sin tax" industries of alcoholic beverages and tobacco; in the view of many potential advertisers, billboards needed a serious image makeover.

There were only two ways to get an advertisement (or "facing") onto a billboard, and both were slow and expensive. First, a unique individual sign could be hand-painted, letter by letter, line by line, color by color. This was labor-intensive and costly. Second, offset lithography could be used to print the facing, which would then be glued by hand onto the billboard. Printing was much cheaper per billboard than painting, but only if a lot of billboards were to display the same image. The printing was done with rare and expensive large-format presses owned by a few specialist shops, which imposed minimums of five hundred or a thousand facings, even if the client wanted to advertise on just a few billboards. Printers effectively dictated high costs to the billboard operators. Even the major operators such as Gannet, Patrick Media, and Metromedia were caught in the squeeze of long print runs, high costs, narrow customer bases, excess capacity, and poor image.

In the early 1990s, multibillionaire John Kluge, CEO of Metromedia, realized that robotic paint-spraying "pens" could be linked to desktop publishing technology—simple personal computers—to make huge, crude, ink-jet printers capable of handling these large-format jobs. Such a printer could digitally spray-paint a 14- by 48-foot billboard facing in three to four hours.

This new application of digital printing would make runs of ten, five, two, or even one facing economically feasible. No expensive labor was needed for hand painting, and because Metromedia would own the equipment, its dependence on offset printers ended. In fact, Kluge believed that Metromedia could use the technology to service the rest of the industry. To do this he set up Metromedia Technologies in Wooster, Ohio, to crank out billboard facings with these new machines. The site in Wooster has since grown to an industry-leading operation, running more than two dozen huge printers working twenty-four hours a day and connected to remote sites in other countries.

Thanks to this new technology, Metromedia ended its dependence on regional and national tobacco and alcoholic-beverage advertising. The natural market shifted toward local advertisers, who consider single urban billboards a major selling point given their low cost and proximate location. The range of customers that could be targeted thus broadened enormously to include local retailers, restaurants, hotels, events, zoos, museums, local political candidates, and radio stations. As billboards grew cost-competitive with newspaper and radio advertising, they became attractive to both smaller and time-sensitive advertisers. Billboards returned to being a widely used marketing tool. Today billboard capacity is up, utilization is up, costs are down, the customer base is broader, and profits are way up. Positioning technology under this business enabled Kluge to lift it to a new, much higher level. To repeat, that lift is the essence of TechnoLeverage.

If technology can drive opportunity in the billboard business, imagine what it can do in other industries.

Money Left on the Table

Our experience at Technology Marketing Group has brought home a clear lesson: Most companies fail to fully exploit their technology. The following examples of technology companies strike me as good evidence of this contention.

- Xerox invented xerography, but Canon now dominates the market for small copiers. Years ago, Xerox made a strategic decision to avoid the low end of the copier market, where its technology was certainly applicable, and to serve only large, high-copy-volume, big-ticket customers. Good-bye low end; hello serious competition from Canon.

- Separately, at its Palo Alto labs, Xerox pioneered the graphical user interface technology that made Apple's Macintosh so distinctive. However, we never saw a PC product with the Xerox name on it (much less "Xerox Inside"). Instead Microsoft, a follower of both Xerox and Apple, muscled in and now makes the real money with its Windows family of software. Money left on the table, by two companies.

- In the 1970s, IBM's leadership in mainframes positioned it for leadership in minicomputers. But like Xerox, IBM thought it was making too much money at the high end of the market to warrant paying attention to the lower end. Digital Equipment Corporation seized that role and grew large and prominent by selling a variety of minicomputers. But then DEC—the company built on making computers smaller—could not understand why anyone would want them smaller still. Both IBM and DEC left money on the table, and Steve Jobs and Steve Wozniak created Apple Computer.

- Under John Sculley, Apple then chose short-term profits over long-term dominance when it decided not to accept open architecture and permit cloning of its superior technology. Without a technology path forward (other than moderately faster chips), and without a true long-term marketing strategy, Apple stopped growing and watched the Intel-equipped PC market sweep past it, grabbing virtually all the money Apple had left on the table.

Something in these highly capable companies' strategies blinded them to the opportunities right in front of them. Why did they not fully exploit what was rightfully theirs? What did their analyses lack?

Companies Need a Strategy for Creating and Capturing Value

The companies mentioned above created value but then failed to capture it. Certainly it wasn't for lack of funds. Even if they had lacked capital, shareholders would have gained if management had just borrowed the money and expanded in the right direction. What was lacking was a strategy of value creation and capture.

What does a strategy look like when it creates value by adding technology? Here are two examples.

- For more than ten years, MBNA and First USA, two of the largest nonbank issuers of credit cards, grew side by side, both achieving fast growth and attractive returns. First USA competed by focusing on marketing. But its chief rival, MBNA, did even better. In addition to its affinity marketing strategy, MBNA focused on applying information technology to customer selection and screening, credit approval, account control, personalized service, and marketing—even direct mail. MBNA strongly outgrew First USA in value. First USA was bought by Banc One, while MBNA continued to apply its strategy and grow independently.

- In oil field services, Schlumberger consistently outperforms chief competitor Halliburton because it relentlessly applies new down-hole technologies to the oil field service business. Ironically, year after year it is Halliburton that develops many of the specific advances that Schlumberger uses to sustain its lead. Halliburton is aggressive in its R&D and creative about product and service development. But Schlumberger has a strategy for getting new ideas—from anywhere it finds them—into the marketplace fast. Halliburton sows; Schlumberger reaps.

In today's economy, companies need a strategy that consciously uses technology for improving business results. A well-executed technology-based strategy enables any company to capture maximum value from the technology it uses. Technology well applied creates market advantage over competitors, even those who may be more technologically advanced.

Basic TechnoLeverage

The term *TechnoLeverage* seeks to convey the image of a business being lifted by technology to a higher level of sales and profits than it could otherwise achieve. The essence of this strategy is to develop every useful application that can be drawn from suitable technologies and to take each of those applications to every profitable market. This is what John Kluge did in the billboard business. He found a useful application drawn from a suitable technology—application of ink-jet and microcomputer technology to printing billboard facings—and took it in all profitable directions—setting up an operation to serve not just Metromedia but the entire industry.

This "pixie dust" aspect of technology can transform a business almost magically, even one as prosaic as billboards or sneakers. By adding useful technology to what you do, you create value and offer something more to your customers. But to add technology to what you do demands that you train yourself to view technology and your customers anew. That's what this book is really about.

Think of How to Solve Problems

It's easy for a business with a successful application to think in terms of pushing product instead of solving problems. Think of where Digital Equipment would be now if it had kept solving the problem of how to make computers smaller instead of narrowing its vision to how to sell more minicomputers. What makes any company successful in the long run is the ability to solve customer problems *repeatedly.* TechnoLeverage calls for keeping the definition of technology in mind—the application of scientific and engineering knowledge to solving practical problems—rather than just pushing product. This is hard to do after you develop a winner, but it's still crucial:

> ► ***TIP NUMBER ONE:* To make money with technology, you must solve practical problems with it.**

This truth unites the technologists, who often pursue technology for its own sake, and business managers, who often fail to perceive its value for the customer. In fact, this truth is so vital that it's the first of ten TechnoLeverage Tips, introduced throughout the book, for carrying out technology-based strategy.

Here is an example of a business that built itself by repeatedly solving customers' problems using a technology-based strategy. Husky Injection Molding, located in Bolton, Ontario, northwest of Toronto, has grown from a nearly bankrupt two-man machine shop to a billion-dollar-a-year manufacturer of plastic molding equipment. When you sit on a one-piece molded plastic chair, chances are that it was made on Husky's equipment.

Husky began as a manufacturer of snowmobiles in the early days of those vehicles, but it quickly failed. The two owners still had machinery and needed work, so they started subcontracting as makers of molds for injection molding of plastic. Their urge to improve led them to reengineer some molds themselves. Using the best technology and techniques they could find, devise, or hire, they developed better molds and were soon selling them to many manufacturers.

They regularly examined how their customers used these molds and asked themselves what to do differently, in light of what the customer was doing. This led Husky to make "hot runners," the heated assemblies of tubes that feed molten plastic into the molds. Making hot runners became business number two.

They soon found that their high-quality molds and hot runners were outperforming the rest of the customers' molding machinery in terms of cycle times, clamping pressure, endurance, torsional strength, and feed capacity. Husky examined the idea of making the rest of the machinery good enough to fully justify running their extremely high-quality molds and was soon in the molding machinery business. Just by moving downstream, they were soon participating in three closely related businesses: molds, hot runners, and molding machinery.

Husky then realized that its customers were making high-quality plastic parts and products, but when the parts came out of the molds they were difficult to handle and sometimes broke. Husky reasoned that if robots removed the parts from the molds, the speed, quality, and reliability of the customers' operations would all increase. This led Husky into manufacturing pick-and-place robots to handle the just-made pieces. More problems solved, resulting in business number four.

Husky later saw that their customers were not organizing their plants to take full advantage of the productive capacity of the much improved equipment. Using the best design concepts and planning techniques available, they started up a service for designing plant layouts that would do exactly that. Sophisticated plant design became business number five.

Today Husky is in five different though technologically related businesses. These add up to a highly competitive, growing, global company at the high-quality end of the injection molding industry. Husky grew by steadily developing new applications that complemented earlier ones. They did this by sequentially seeking problems they could solve. In the process, Husky built itself from two desolate machinists into one of the world's top players in this specialized, competitive business.

Think About Customers in New Ways

Although Husky succeeded by perceiving how its customers could benefit from technological improvements, there are other ways to go about leveraging what you do. Admittedly, the practice of finding out what customers say they want and then supplying it—and not supplying it if they say they don't want it—generally works. But simply responding habitually to the customers' expressed wants can get you into trouble where technology is concerned.

Consider the development of automatic teller machines. In the 1970s, Citibank's research revealed that its retail banking customers would absolutely not use ATMs. In survey after survey, customers said they would not trust their money to machines, they would miss dealing with tellers, and they would find the perceived lack of security unacceptable. Customers told Citibank, loudly and clearly, not to develop and adopt this technology.

With its commitment to leadership and its need to control costs, Citibank spent $250 to $300 million to develop its network of ATMs for this skeptical market. As we all know, ATMs eventually met with phenomenal market success, for Citibank and the entire industry. However, because Citibank led the customer to the technology, the bank established the first and widest system of ATMs in the New York market. During the late 1970s and for much of the 1980s, Citibank's retail banking business benefited from larger consumer deposits gathered at lower cost than their competitors.

To take another example, in the late 1970s who knew that there was a multibillion-dollar-a-year need for overnight package delivery?

Before the personal computer revolution, who imagined that senior managers would learn to type—that is, use a keyboard—just as their secretaries did?

In the early 1970s, how many doctors could have told medical equipment manufacturers to use fiber-optic cable to fashion laparoscopic devices to see into the human body?

Because new technology is often broadly applicable but poorly comprehended (as with medical fiber optics), and because the benefits of a technology for the user can be so poorly perceived (as with ATMs), customers cannot accurately tell us how to apply technology to our business in a way that offers them more. Furthermore, it's not their job. It's our job. Customers look to companies for innovation. When applying technology, you must often lead your customer. When you have an application that solves a problem, one that you feel you can use to lead the customer, act quickly and get it out there.

Think of How to Get It Out There

Particularly in today's fast-moving marketplace, it's best to get an innovative new application to customers sooner rather than later. Focusing on speed-to-market helps you overcome one common obstacle facing technology—the don't-release-it-till-it's-perfect syndrome. Although users of flawed software releases may scoff at the notion, most technologists—most scientists and engineers—enjoy the pursuit of perfection in their work. They frequently forget that there are interested customers out there waiting for something that is simply *better.* The new product doesn't have to be perfect as long as it is better. I urge my clients to get it out the door, and then keep working on perfecting it.

Getting it out to customers fast brings in valuable information from those early customers, especially regarding performance requirements, ease of use, and compatibility with existing operations. Customers using early versions of a product are likely to go beyond commenting on simple performance attributes such as speed, strength, flexibility, and durability. Early customers are normally willing to share their problems with you. Then you can mull over several questions. Do they understand it? Can they use it? Do they like it? Are they excited by it? Does it help them in their business? Would they buy it?

From 1965 to 1995, Eastman Kodak was an outfit that failed to "get it out." Kodak had an enormous R&D budget; in the lab it vigorously pursued improvements in digital cameras, digital imaging, and various image storage and reproduction applications, but none were perfect. Kodak reasoned that these innovations might hurt its base business if they were released. Because they weren't perfect, they remained in the lab. Today, spurred by consumer interest in electronic imaging, by competitor Fuji's ascent, and by the forceful leadership of CEO George Fisher, some "new stuff" is coming to market, signaling the company's efforts to awaken from its TechnoSlumber.

Not all companies are afraid to bring imperfect advances to market. Lack of such fears gave Hewlett-Packard and Hammermill huge leads in the ink-jet paper market. Both entered the business early and used technological advances combined with marketing savvy and brand-name clout to gain these leads.

Ink-jet printing employs water-based dyes and coated paper. For a sharp image, the paper must trap the dye on the coated surface and keep it there. The paper beneath must be formulated so that it pulls the water in the dye down from the coated surface and into the paper, from which it evaporates, thereby drying the image.

Ink-jet coating on paper is a tricky technology. But the difficulty didn't stop HP and Hammermill from going to market as soon as they had commercially acceptable formulations—as opposed to technologically perfect ones. HP has no paper mills, so it worked closely with other companies' mills under contract. Hammermill, a division of International Paper, worked internally. Both companies endured early criticism, price resistance, and customer complaints about the paper. However, they both kept improving their products, a process that I suspect continues to this day. Thanks to getting it out fast, the two companies now dominate the ink-jet paper business.

Just these three simple practices—solving problems, leading the customer, and getting applications out quickly—can, if consistently executed, raise your business to new heights. Yet the path of a new application, even a successful one, does not inevitably lead to everlasting success—quite the contrary, in fact.

The Path to Success—or Failure

Business history and the marketplace are littered with the remains of once successful products that were superseded by better technologies or became commonplace items that added minimal value. This includes everything from erstwhile giants like passenger railroads and network television to items such as the typewriter and the adding machine, or more recent offerings such as the 5.25-inch floppy disc and Visicalc, the first mass-market spreadsheet for personal computers, introduced around 1978.

Where will a successful new application lead? What market dynamics determine the future of your product and company?

Figure 1.1: The Technology Applications Spectrum (TAS)

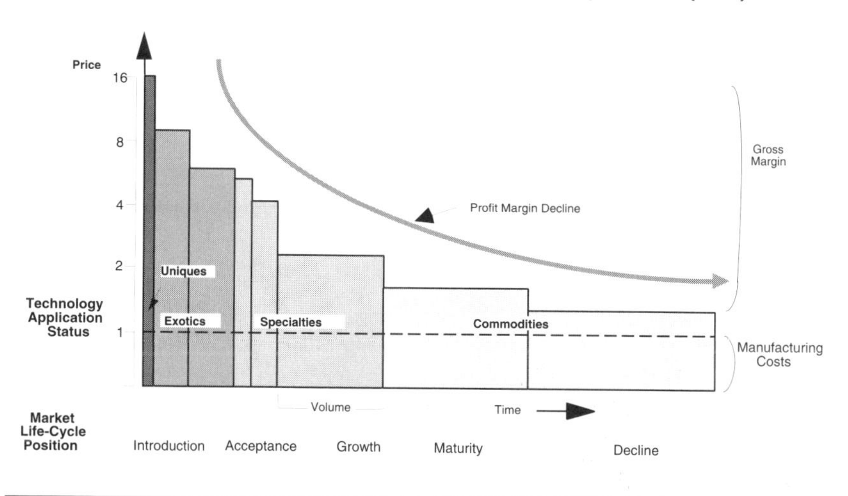

What kind of money can you make—and *when* can you make it? What competitive threats await you?

There is a superb tool for attacking these and similar questions. It's a road map, called the technology applications spectrum, that grew out of our work with clients and our observations of scores of client products tracked over decades of existence or extinction. The technology applications spectrum, or TAS (Figure 1.1), is a graphic that tracks the progress of an application of a technology over time. It was originally developed to help companies that owned technological capabilities find uses for those technologies.

Before we introduce the graphic, it might be useful to distinguish among the terms *technologies*, *applications*, and *products*. For example, Kaman Corporation, a Connecticut manufacturer of military helicopters, has vibration-management *technology*. It uses this technology for two completely different *applications*: Kaman reduces vibration resonance in the rotors of its helicopters, but it increases vibration resonance in a line of guitars made by one of its divisions. These applications take form in two *products*, Kaman's K-Max helicopters and its Ovation guitars.

As the TAS shows, a successful application (helicopters, guitars, and so on) inevitably follows a path from innovation to maturity. This path traces the unavoidable decline in gross margin that occurs as an application moves from being an innovative use of a technology to becoming a commonplace item. The TAS has significant implications for any company using technology for strategic advantage, so let's spend a moment with Figure 1.1.

The TAS depicts four stages, or zones, classifying an application as a unique, an exotic, a specialty, or a commodity. These zones correspond both to the breadth of market acceptance that the application has achieved and to the scale of production:

- A unique application is a first-of-its-kind solution, an application of technology that offers special characteristics or specifications not available elsewhere. It is often made to order, such as a telecommunications satellite or a factory system. Or unique can mean an early-stage solution just introduced and destined for wider use.

- An exotic is an application still in the early stages of introduction and not yet mass-produced in any sense of the term. However, the application has won wider acceptance than a unique one, and if it keeps on proliferating, it will become a specialty.

- A specialty is a mass-produced category (note the third, larger box on the diagram within the specialty zone, which designates the case of higher volume) that has moved to wide market acceptance and significant growth. A maker of a specialty application can stay in that zone by repeatedly improving the product through technology.

- A commodity is a truly mass-produced item that has penetrated the mass market and reached maturity. As in the usual definition of a commodity, these products tend to be largely indistinguishable from one another. Their distinctive value-added is minimal, even though they may be extremely useful, as is aspirin.

This last point about value-added brings us to the crux of the situation illustrated by the TAS. As an application moves from one zone to the next, the profitability as measured by gross margin drains out of it. For reasons that are explained in detail in Chapter 7, gross margin is the best measure of the value added by the application for the customer. Uniques and exotics command large margins because they represent either the only solution or an unusual solution to an expensive problem. Competition facing uniques and exotics is minimal because the product has not won wide enough acceptance to draw competitors into the business. The provider of this rare solution to an expensive problem can add large amounts of value. Suppliers of exotics have willing buyers and few competitors, so their gross margins are high.

However, as the application continues to win broader acceptance, other suppliers see the attractive margins in the business. Competitors begin to enter the market, and value is siphoned off as customers learn how to make it themselves or buy substitutes. The game changes to a mass market, which requires mass production. Standardization and cost reduction become priorities. The presence of all those competitors gives customers a choice of suppliers, enabling them to reward only suppliers that drive the price down. This is how margins dwindle. It's how specialty applications move into the commodity zone.

If an application is out there steaming along and gaining acceptance, a temporary short-term decrease in gross margin over several months caused by temporarily higher raw material costs should not send management into a swoon. If costs are the problem, the company needs to control them, and things will improve—returning gross margins to where they were.

However (and I stress this because it is so often ignored), over the long term the profitability of an application necessarily declines. Again, this results from downward pressure on margins as competitors enter and as purchasing agents seek substitutes, demand concessions, and play suppliers against one another. At that point, the application's full maturity has arrived; the remaining value-added is generated by production efficiencies and better distribution. Nearly all of the value originally added by technology has been wrung out of the application.

What Do You Do Next?

Once the application reaches maturity, most managements find themselves facing a costly share battle that requires them to invest more and more money to chase inherently diminishing margins. This is the problem with market- and operations-focused strategies, which tell us to fight for share. The effect is to ride the curve downward into the Land of Low Margins. Technology strategy provides an extremely attractive alternative.

For their part, traditional marketing tactics involve the five P's (product, price, package, place, and promotion). They seldom focus on what made the market grow in the first place: application of the technology. Marketing's focus on these tactics rather than on technology led Dr. Amar Bose, chairman of Bose Corp., the leader in high-fidelity speakers, to tell *The Wall Street Journal*, "Marketing people's perfect product is something that has one more knob and is one dollar cheaper." Bose Corp. succeeds by steadily developing new applications for its technologies, which are acoustic science and sound-transmission hardware.

On the other hand, operational strategies attempt to rescue declining profitability by lowering the cost per unit. Recent efforts include reengineering, total quality management, continuous improvement, just-in-time practices, factory automation, and similar measures, all of which enjoyed their greatest popularity in (Surprise!) established companies in mature industries. Among the older and more primitive weapons in the operational arsenal are pitting suppliers against one another, reducing employee benefits, tightly controlling wage increases, and moving production to lower-cost (often foreign) sites. With such tactics, the opportunities for creating new value for customers are slight.

Each of these marketing and operational measures has its place. Yet to the extent that they displace the growth strategy of using technology to solve practical problems and create value, they divert management's attention.

Along the TAS there is room for various styles of play. Not every company should follow its application down the TAS curve labeled "profit margin decline" while fighting for higher share and lower costs. Each zone is off limits to some companies and home sweet home to others. For example, Cargill, the giant agricultural products processor, does very well in the commodity zone of its industry's curve. By contrast, Intel decided to exit the commodity business of memory chips and concentrate on the exotic and specialty applications for its microprocessors. Intel has repeatedly moved back up the TAS by bringing out faster, more powerful, more capable chips. There is room in the zones of the TAS for all kinds of players, provided each reads the signals properly, consciously stakes out a position, and runs the right type of business for its zone. Chapter 9 discusses the different styles organizations can use for different zones.

Technology-based strategy, TechnoLeverage, represents the surest path to success in business today. TechnoLeverage provides workable solutions and practical tactics that enable you to leverage the growth and profitability of your company to new, perhaps undreamed-of, levels. Witness the John Kluges and Hewlett-Packards of the world. The sad results of lacking a technology-based strategy have been demonstrated by the many one-trick ponies who stopped solving problems after introducing a single successful product.

The dynamics of our worldwide marketplace today call for technology-based corporate strategy because technology drives opportunities, customers demand ever increasing value, and competitors instantly copy whatever makes money. Selling undifferentiated products in the Land of Low Margins is not for everyone. Introducing advanced performance products up in the High-Margin Mountains may be more to your liking.

Review and Preview

In this first chapter, we've seen that technology-based strategy can transform any business by opening up opportunities for higher profits and more rapid growth. Technology strategy is a productive alternative to operational and marketing strategies, both of which seldom add the kind of value for customers or companies that technology can. This is because today technology—broadly defined—drives business opportunities.

We've seen several examples of companies, including some in low-tech industries, that use technology to gain leverage. We have also covered some new ways of thinking about solving problems, about customers, and about how quickly a new application should be released. We've identified the first TechnoLeverage Tip: *To make money with technology, you must solve practical problems with it.*

The TAS illustrates the path of an application over time, from unique to exotic to specialty to commodity, as the gross margin inevitably declines. This pattern holds serious implications for every company. How a company negotiates or forestalls movement from one application zone to the next dictates its long-term success or failure.

In Chapter 2 we see that some companies aggressively use technology to master these zones. Despite the risks and uncertainties inherent in technology, they consistently earn above-average returns over time. To see how they do this, let's examine the mechanics of TechnoLeverage.

CHAPTER 2

The Legendary Lever

Lifting Your Business With Technology

Leverage works wonders. Whenever we use it to lift something heavy, solve a thorny problem, or do more with less, leverage provides an advantage. By using technology to gain leverage in business, we achieve the difficult goal of lifting corporate profitability and shareholder value with less than brute force.

TechnoLeverage purposefully applies technology as a lever to business. It can be *any* business. For a tasty example, try this: Go to a large grocery store and buy a jar of Smucker's Goober Grape. The eighteen-ounce jar costs you about $2.39. It's about the cheapest strategy tool you can buy.

Goober Grape is easy to recognize, with its pink and purple lid and label and its cartoon drawings of a happy-go-lucky peanut and goofy-looking grape. This product is, as it says on the jar, "Peanut Butter & Grape Jelly Stripes." Smucker's, which makes various peanut butters and a huge line of jams and jellies, took an idea that had been around for years—peanut butter and jelly—and, with a little help from technology, created a new product with added value and intense appeal for the peanut-butter-and-jelly set. Basically, Smucker's takes a jar of peanut butter and, using a six-pronged insertion device, shoots jets of Smucker's grape jelly down the inner sides of the jar to create the stripes and a ready-to-spread PB&J filling. They charge more for one ounce of Goober Grape than they charge for an ounce of jam and an ounce of peanut butter combined.

Put your new jar of Goober Grape on your desk and keep it there. It will reinforce your efforts within your organization. If someone challenges you, saying your company's products can't be spruced up, you have Exhibit A right on your desk. "Here is a product," you can say, "made up of two commodity elements: peanut butter and grape jelly. Yet this company creatively applied a bit of technology to those two commodities to make a distinctive and high-value product. We can do something like that, too."

The managers at Smucker's are not technologists. They are jam-and-jelliologists with a quality-oriented national brand. Even though their competitive advantage does not spring from technological wizardry, they accomplished TechnoLeverage. By innovative use of technology (quite possibly someone else's technology), they created a distinctive product. This creativity is a key element in TechnoLeverage: applying technology of some kind to any product to differentiate it and give it market and financial advantages.

The fact is that TechnoLeverage—wedging technology under a business in order to lift it up financially—requires thought and change. To benefit from this leverage, most management teams must think in new ways and do new things. Habitual thinking precludes the practices that create TechnoLeverage. Achieving outstanding business results with technology begins with understanding the normal barriers to improving corporate financial performance. Once you understand these barriers, you can begin to overcome them.

Why TechnoLeverage Is Necessary

If TechnoLeverage provides "lift" by applying technology to a business, where's the "drag" that resists the lift? What retards earnings growth? Let's examine the sources of resistance.

Like all businesses, yours is trying to grow in size and profitability. Most companies face major impediments to growth. I've shown some of the common ones in Figure 2.1.

Like most companies, you are boxed in, as the figure suggests. Your products or services are comparable to other producers' offerings. You can't raise prices because your direct competitors undercut

Figure 2.1: Forces That Resist Efforts to Improve Your Business

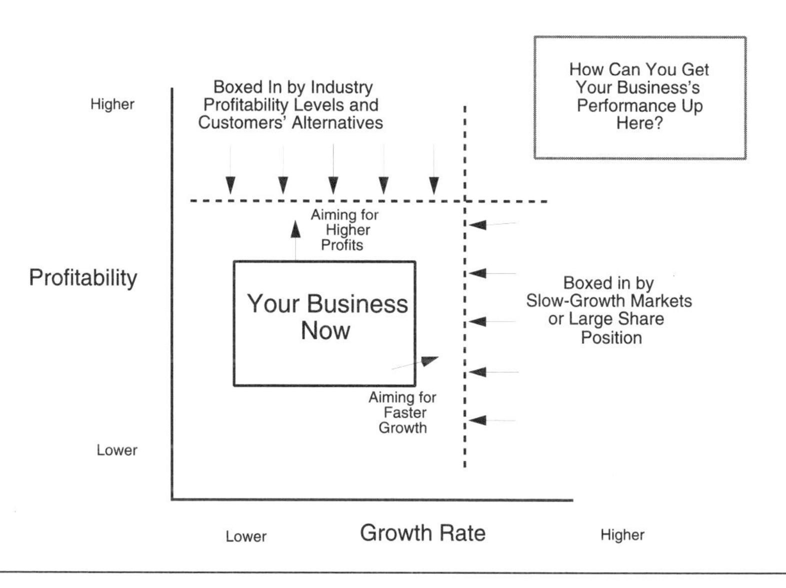

you and your customers have the power to choose substitutes. Your market is not growing as fast as you want earnings to grow. If you're in a technical or business-to-business market, you might have a large share of the market, and the remaining players won't give up share without an expensive fight. The question is how to break out of these confining forces to keep your company growing.

These impediments present almost intractable challenges. Your company can seldom control its customers, can almost never control its competitors, and usually cannot drive others out of the market. You cannot control economic growth, interest rates, tax rates, the value of the dollar, or suppliers' prices. Still, you need not be immobilized.

In the first chapter, I said we'd play a different game. We're going to use technology to lift ourselves out of this box. However, before we start a different game, let's take a look at the standard game.

Moving Outside the Box

Companies generally compete on three broad dimensions: price, quality, and service. To be competitive with other players in your industry, you can give your customers lower price, higher quality, or better service. (*Service* here includes such things as good distribution, fast delivery, postpurchase support, and attractive warranties.) Using the tools any industry normally employs, it is hard to deliver distinctive value on all three of these dimensions at the same time and still make money; for example, high quality and high levels of service at a low price simply aren't normally compatible. Therefore, companies avoid competing on all three dimensions at once. Thus we have the salesman's classic offer to the demanding customer: "Price, delivery date, or quality. Pick any two." The salesperson—and the company—admit that they can deliver on only two of these dimensions at once.

I realize that most companies would like to deliver all three, but normally it can't be done. Instead, a company must choose which dimensions it competes on. In New York, Moe Ginsburg's famous-label discount men's clothier competes on price and quality, but its service consists of a warehouse environment starting three floors up on Fifth Avenue, without a tailor. Federal Express provides quality and service, but not for the price of a postage stamp. A manager from a top-three software company recently admitted that they've stopped providing personal telephone support on established products, because "People are just crazy to expect lifetime support on a $299 product. It can't be done!"

Now let's add technology to your company-in-the-box. Technology enables you to compete more effectively on any or all of the standard dimensions of price, quality, or service. Technology also allows competition on an entirely new dimension: uniqueness.

First, technology enables you to compete more effectively, sometimes much more so, on the dimensions of price or quality or service. This is why you can position technology under any business. For instance, technology may help you do what you already do more cheaply so that you can lower the price. Or you can use technology to increase your level of quality, for example by using it to improve purity or avoid defects in the manufacturing process. Or you can use technology to improve your distribution, or deliver better, faster, or more customized service.

When Nike applies technology to make its shoe soles grip better or last longer, it's using technology to improve its product quality. When FedEx makes a Web-based tracking system available to customers, it is competing on service. (Before it did so, customers could telephone, but the Web is easier for many of them, and cheaper for FedEx.) These are instances of using technology to improve routine competitiveness.

Second, as I've hinted, technology enables innovative businesses to break free of the three dimensions by competing on the new dimension of uniqueness. How? By creating something entirely new, radically different, overwhelmingly better. When Xerox introduced the plain-paper copier, when Apple created a user-friendly personal computer, when VisiCalc created the electronic spreadsheet, when Sony invented the Walkman, they each devised something totally new. The products did not represent incremental improvements in price, quality or service. Instead, each represented something completely new. These creations spawned entirely new industries.

The beauty of technology as a business tool is its capacity to radically alter the competitive landscape. By applying technology, you can develop breakthrough products or start new industries. In this way, technology adds a fourth dimension of uniqueness to the standard three dimensions on which companies traditionally compete. This is the basis for TechnoLeverage Tip Number Two:

> ▶ ***TIP NUMBER TWO:* Use technology to create differentiation on which to build high-margin, high-growth new business.**

As shown in Figure 2.2, technology alters the terms of competition so that you are not restricted to competing on price, quality, or service. For example, if a competitor cuts prices, increases quality, or improves service, you can try to match it if you wish. The competitor might then match your improvement along the same dimension, setting off a leapfrog series of improvements—or worse, a price war. But if you use technology to create a unique offering that supersedes the competing product, you gain a substantial advantage. Think about the genius who devised a shoe with plastic wheels one behind the other on the sole to make an inline roller skate—the birth of Rollerblade, Inc. A little bearing technology, a little plastic technology, a little creativity, and presto: a new industry, a new sport, a new means of urban travel, a new way to meet the opposite sex.

Figure 2.2: Technology: A Tool to Get Out of the Box

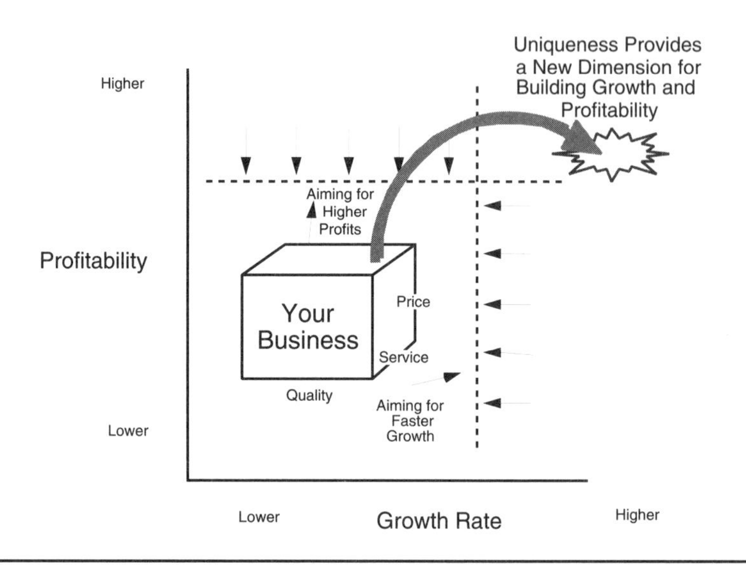

Of course, not every new idea for technology leads to success. Someone's new product can in turn be superseded by someone else's new product. This can occur even if you correctly gauge the market's desires and judge that your product fulfills those desires.

When Polaroid saw one of its products immediately superseded, only a much broader view of the market and the competition could have saved it. In the early 1970s, Polaroid was preparing a product called Polavision for market release. The company saw Polavision as the next generation in home movies and believed it would replace regular 8mm film as the standard. The product, which actually got to market for several weeks, amounted to "instant-movie technology." Accompanying the camera was a little machine that the user put the film through after it was shot. The machine stripped off the carrier sheet that held the famous Polaroid instant photochemistry, leaving a transparent strip with photographic images ready to be viewed in a movie projector.

Polaroid had gauged the market need correctly. Consumers did want a home-movie technology with better resolution and ease of use than 8mm. Certainly, Polaroid's new offering was unique. However, technological uniqueness does not guarantee success or high returns. Unfortunately for Polaroid, Polavision was introduced just months before the first video cameras appeared and became the standard. In the eyes of consumers, cheap, long-running, one-step videotape instantly superseded expensive, short segments of two-step film. Polavision was immediately beaten by the newer video technology. It had been around for years, but since that was not Polaroid's area of expertise the company had not focused on it. If they had looked at new developments in video, they might have avoided a costly and embarrassing disappointment.

If your product can be superseded by another technology, you face a serious threat to your business. Just as videotape vanquished Polavision, the calculator doomed the adding machine, and the personal computer replaced the dedicated word processor, which had outmoded the electric typewriter. Clearly, new applications of technology have the potential to move you to new competitive heights—or they can fizzle.

What can you do about this risk? One tactical step is to include *scanning* in your planning, as we see in Chapter 3. Scanning the business environment helps you develop a broad view of the market and potential competitors. But shouldn't a well-thought-out technology strategy also reduce the risk and provide mostly rewards? If it does, how big are the rewards?

The Returns on Technology, for All the World to See

Working with all kinds of technology originators and technology users year in and year out, I began to wonder exactly how big the returns on technology actually were. Everyone can see the Silicon Valley example of wild financial success, so high returns are clearly possible. But, I asked myself, how predictable are those returns? What about companies that just make specialty chemicals, or rare alloys? What about companies that have a little technology edge, not a big one? If we can see the risks, can we measure the returns?

I reasoned that if technology creates uniqueness and thereby establishes a separate competitive dimension, we should be able to analyze technology separately. We should be able to examine it in isolation, understand how it operates, and grasp how to manage it. We should be able to measure the returns it yields. In 1994, Technology Marketing Group attempted to research technology in this way—with unexpected results.

I knew of no research that specifically examined the returns of technology-based businesses, broadly defined as companies whose value-added comes primarily from technology. It seemed useful to study a wide range of enterprises working with all sorts of advanced technology. At TMG we decided to research a full range of technology-based industries: ceramics, adhesives, specialty materials, robotics, chemicals, instruments, measuring devices, and photo-optics. The goal of the research was to isolate the returns on technology.

What We Did

Our research began with the reported performance of public companies making and selling a variety of technologies. We established two sample groups so that we could compare them. One group represented "technology" companies; the other group represented "nontechnology" businesses. The first sample consisted of one hundred publicly traded technology companies randomly chosen from the CorpTech *Corporate Technology Directory*, a well-known database of businesses operating in twenty-three technology sectors. The second sample comprised one hundred publicly traded companies randomly chosen from *Walker's Corporate Directory of Public Firms*, found in most public libraries. In preparing the nontechnology sample from *Walker's*, we excluded any company that was also listed in CorpTech because such businesses are all, by definition, "technology firms." What the nontechnology sample contained were companies in familiar businesses, such as the retail, restaurant, food processing, and consumer packaged-goods industries.

All companies in the study had at least $25 million in annual revenue as reported in their 1993 fiscal year annual reports. In 1996, we replicated this study using the same companies' 1995 annual reports and found essentially the same results. I believe we would find the same results again today.

To analyze these businesses, we:

- Compared the two samples of companies on nine simple standard measures, including sales, growth rates, employment, and total assets

- Compared the two samples on twenty widely used computed performance measures that combine two or more of the simple measures, including, for example, gross margin (net revenues less manufacturing costs, divided by revenues), net margin (net revenues less all operating costs, divided by revenues), percentage growth in revenue, research and development (R&D) expenditures to profitability, revenue per employee, and profit per employee

- Analyzed these data using standard statistical tools, including regression, correlation, and factor analysis to isolate the key issues regarding performance

We repeatedly compared these two groups along different measures, trying to find the returns on technology. We sought to measure the financial rewards we felt they must be enjoying, rewards that accrued to the uniqueness that technology provided them.

What We Found

The two groups closely resembled one another by most measures, which was puzzling in itself. However, in one area—return on assets (ROA)—the two samples clearly differed. For someone like me, who believes in technology as a business tool and who fully expected to see positive returns on technology, the results were surprising. Our major finding was that the return on assets for the technology companies was actually *lower* than the return on assets for the nontechnology ones. This seemed to indicate that the returns on technology were actually slightly negative. Such a result was completely unexpected.

The technology companies produced a median return on assets of 11 percent, while the median ROA of the nontechnology companies was 12 percent; this is depicted in Figure 2.3. We had expected the results to be the opposite.

Technology companies do not have accounting entries for many of their most valuable technological assets—"off-the-books" assets such as unique product offerings, technical leadership, the advanced knowledge of their scientific staff people, exceptional technical manufacturing know-how, and their ability to devise, make, and sell solutions to previously unsolved problems. Although these assets are off the books, technology companies should be making money from them. Unlike the nontechnology companies, they have a technological means of adding value, so their return on book assets should be higher. Thus we were taken aback to find that the returns of the technology companies were in fact lower than the returns of the nontechnology ones.

Not only did there appear to be no overall financial advantage from technology, but there seemed to be an explicit penalty. Technology adds complexity, uncertainty, and risk to the task of managing. Perhaps these businesses are not being repaid for their trouble. Technology companies require more investment in R&D. Maybe these assets aren't earning their keep.

Technology companies operate in a less stable environment and face sophisticated, demanding customers. They employ and manage expensive scientific, technical, and engineering professionals. In a sense, this study indicated that many participants in underperforming technology businesses—and their investors—should consider cashing out and using the proceeds to get into the restaurant, retail, food, or consumer packaged-goods business. Why knock yourself out for lower returns?

Our second key finding was more nitty-gritty but just as useful. The technology companies displayed a much wider dispersion of return on assets than the nontechnology ones. As shown in Figure 2.3, the returns of the nontechnology businesses were much more compactly distributed than those in technology. This implies that there is much greater consensus in the business world on how to manage nontechnology businesses—retail stores, restaurants, and wholesalers—than there is on how to manage technological ones.

Figure 2.3: Are There Positive Returns on Technology?

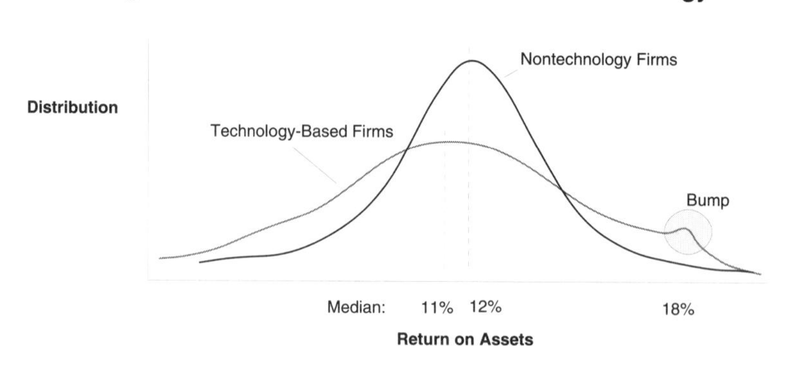

The tighter distribution of returns for nontechnology companies says that as a society, we've figured out how to run the businesses we've had in our economy for decades. We've had time to learn what works and what doesn't in the nontechnology businesses. The question is: How do you manage technology businesses?

A Hopeful Finding at Last

By pointing out the problem of lower returns, this research supported the observation that running a technology business can be a chancy proposition. Investing in them in the aggregate might be even more hazardous. However, we uncovered some good news as well. It has to do with the bump at the high end of the ROA curve for the technology companies in Figure 2.3. It represents an anomaly in the findings about these enterprises and points the way out of this high-tech, low-return problem. These companies support what I have seen in twenty years of working with technology-based businesses, namely, that some technology companies are consistently very profitable and extremely well run.

As shown in Figure 2.3, the bump at the high end of the ROA curve represents a small cluster of companies with an ROA of about 18 percent. These few were achieving what one would expect most technology businesses to achieve: extraordinary and consistent returns far exceeding the median return for nontechnology companies. This cluster at around 18 percent ROA earns the kind of returns that say, "Take advantage of technology. As a businessperson, use it; as an investor, pursue it."

When we repeated the 1993 study with these businesses two years later, we expanded the technology sample by two hundred. As noted, using 1995 fiscal year data, we found essentially the same distribution for return on assets with this larger, later group. Our sturdy band of 1993 standouts still stood out. Clearly, they had something like a recipe. Though their performance may change by the time this book comes off the press, at this point we can say that they, and companies like them, seem to know what they're doing with technology.

With returns that far exceed the median of the rest of the technology sample and of the nontechnology sample, these companies provide a field guide to knowledgeable technology users. Some of their names are familiar, some are not:

- Abbott Labs (medical supplies)
- Analysts International (systems analysis and computer consulting)
- Biomet (medical prosthesis and orthopedic replacements)
- Cognex (machine vision systems)
- Empi (biomedical engineering)
- Franklin Electronic Publishers (publishing platforms for electronic books)
- Wellfleet, later merged into Bay Networks (local area network solutions)
- X-Rite (precise color measurement systems and equipment)

In our work for clients we often act as scouts, traveling in uncharted territory and looking for obscure clues. The high-return companies, though only a small portion of our sample, were earning the returns that we believed most technology enterprises should. Moreover, they weren't just having a great year; they remained ROA leaders. They acted as trail markers for us, yielding a clear enough pattern of values and behavior to solidify a model for managing technology. Interestingly enough, one can find other businesses following this same pattern, and they usually have the same attractive returns. This model works as well for technology-intensive businesses as for nontechnology ones that need a boost in growth and earnings.

What's the TechnoLeverage Formula?

In reviewing the organizational values and management practices of the cluster of high-return technology companies in our sample and then proving out the model on others, we found several recurring themes. These businesses all seemed to do similar things most of the time. Technology companies in dissimilar industries were using the same fundamental approach to business. This approach is clearly usable in many fields. Here is the formula for what TechnoLeverage organizations do:

- *They concentrate on their core technological skills* as a primary source of value-added.
- *They invest consistently to enhance those skills* over time.
- *They search widely for new markets* that value their capabilities and for new applications that employ their technologies.
- *They target markets with the highest growth and profitability,* especially markets where the company's technology provides an enabling capability.
- *They try to dominate or lead* in their area of influence, whether that area is a market, an application, or a technology.
- *They work toward selling whole products and complete systems,* and they try to constantly broaden the capabilities of their applications and products.
- *They are aggressive and sophisticated about pricing.* They price based on value rather than cost, which means they price high, especially in the early stages of the life of an application. As the value they add declines, however, they lower their prices accordingly.
- When the time comes, *they abandon low-profit, low-growth applications and markets.* They are smart about getting out of, or avoiding, situations where they provide no unique high-value capability, or where pricing is determined by others.

This set of practices forms the seed of the technology-based strategy I call TechnoLeverage. Refined into a corporate strategy,

TechnoLeverage can guide all managers toward higher revenues and profits. Using technology-based strategy, a company can get the added value, the leverage, the something extra that technology should be yielding to all who use it as a business tool. If some companies have consistently done it, others can too.

Good strategy does not happen by itself. This research clearly shows that being a technology-based company does not guarantee the benefits of TechnoLeverage. You must know how to manage these companies. Simply grafting technology onto a low-tech business does not automatically produce results. As the data show, ordinary involvement with technology holds more pitfalls and offers returns lower than those regular businesses earn. Hard work, a clear model, and perseverance are necessary for technology-based strategy to succeed.

Raising ROA

A number of the eight practices we identified as elements of TechnoLeverage directly affect underlying return on assets. By way of background, at the simplest mathematical level, if ROA is too low, either the assets are too large for the return they earn, or the return is too small for the asset base that's been assembled. These diagnoses are simply two perspectives on the same ratio. Each one indicates a different course of action: Either get rid of some assets or put what you have to work immediately to increase the returns.

Companies with a true technology-based strategy do both. They generally reduce the amount of assets their operations require, while at the same time they try to enrich the returns that the remaining assets earn. This happens to fit exactly with what popular economic value-creation programs such as EVA try to accomplish.

Here's an example from the paper business that illustrates the last point on the list above. Arnold Nemirow has been elected twice as Papermaker of the Year. Nemirow is one of the best at TechnoLeverage; I watch his moves carefully because he plays so smoothly. In 1996, he became president of Bowater, Inc., a major newsprint manufacturer based in South Carolina. He made no major public moves for a year or so. Then one day Bowater announced that it would sell its timberlands to deploy the assets into more-productive activities. Sure enough, Nemirow's first move was to withdraw the assets his new business was deploying in a low-return activity (others can make an attractive return managing timberlands, but this man decided his company wasn't one of them) and redeploy them into more profitable papermaking endeavors. Good executives like this are unsentimental about heritage businesses, conventional industry structure, and product mix. They abandon low-potential opportunities gracefully.

In the context of technology-based strategy, to manage assets properly you must invest money and other resources regularly to apply any technology that enables you to add value. Then you must search widely for new applications that can use those technologies profitably and for new markets that value them highly. That way, your total asset base stands the best chance of being fully utilized and producing the greatest return. *The goal is to utilize all your assets.*

In Chapter 7 we examine in detail the financial measures by which to steer the company. We also explore the contribution that financial managers make in this overall effort. But the simple practice of monitoring ROA and seeking the kind, amount, and balance of assets that generates maximum returns does a lot to keep the company on the right path. Monitoring ROA leads to the heart of the leverage issue.

The Mechanics of the Machine

We have discussed the results of research into technology companies' financial performance and seen how certain companies achieve extraordinary business results. Now let's get back to the idea of the lever and explore the mechanics of TechnoLeverage. Again, the leverage we seek comes from positioning technology under the business for the sought-after "lift" in financial performance.

A Picture of TechnoLeverage

If TechnoLeverage is the lift that results from applying technology to a business, how does it work in practice? Figure 2.4 shows one way to think of TechnoLeverage.

The fulcrum in the figure represents the company's assets, while the shaded square represents the company's revenues, positioned at a specific growth rate and profitability. The vertical arrow in the upper right represents the growth rate, size, and profitability of the *additional* business the technology makes possible. The lever itself represents gross margin. Essentially, you bring technology to bear on the company's profitability, size, and growth—its overall performance—through the relatively high gross margins and fast growth of the new business.

Here is the essence of the kind of leverage that our "bump companies" revealed: The portion of the lever to the right of the fulcrum represents the gross margin on the *new* business that technology wins for the company. The portion of the lever to the left of the fulcrum represents the margins on the older, established, relatively mature businesses of the company. Since the new business is markedly higher-margin, the portion of the lever that represents this business is longer. The new, higher-margin business leverages *all* of the company's performance measures, thus allowing you to overcome the forces opposing growth.

We can take this example one step further and show how this works mathematically in very simplified form.

Table 2.1 shows the results of a hypothetical company or division of moderate size, earning average returns, establishing an

Figure 2.4: Applying TechnoLeverage

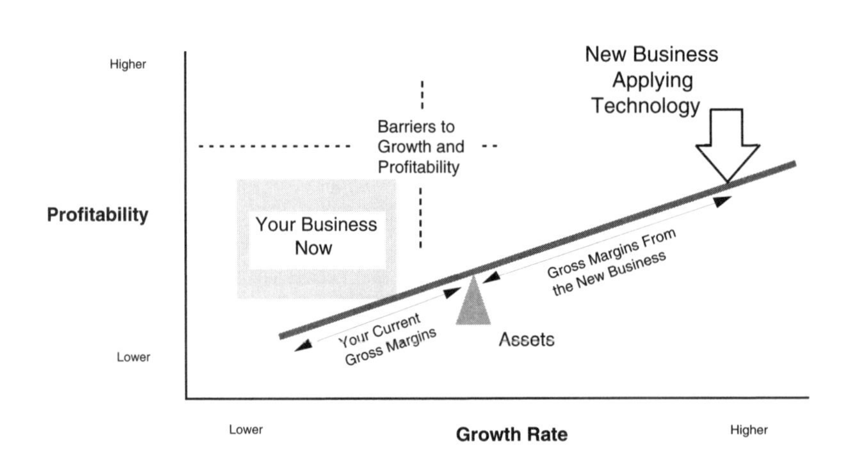

Table 2.1: The Financial Impact of TechnoLeverage

	Base business	Expansion business	New result	Change
Revenue	$200 MM	$20 MM	$220 MM	+10%
Net profits	$10 MM	$6 MM	$16 MM	+60%
Return on sales	5%	30%	7.3%	+48%

expansion business based on its use of its technology in a new business area. Without making any changes in its base business (which, as we have seen in Figure 2.1, is hemmed in), the company found an opportunity for a new business only one-tenth the size of the base business. But the higher margins of the new business dramatically improve overall corporate performance, expanding net profits by 60 percent. Because a relatively small business unit makes this kind of attractive improvement in the performance of a much larger parent company, I call this approach TechnoLeverage.

As I pointed out in Chapter 1, it is natural for applications to mature, for markets to become saturated, and for margins to decline. There are irresistible forces—market, technological, customer, competitive, and economic forces—constantly putting downward pressure on companies' margins, constantly pushing each company's resting point on the lever in Figure 2.4 to the left, toward lower performance. Management's eternal job is to use the company's resources to resist these forces. Management can use technology to repeatedly move the company into new, higher value-added applications and new, higher-margin markets. As a manager, you must find new business opportunities that move the large arrow's pressure point on the lever further to the right. Technology helps you average your margins upward.

Figure 2.4 and Table 2.1 reflect the underlying mathematics, which says that the primary effect of TechnoLeverage is to use high-margin new business (even though it usually starts off small) to significantly boost the returns of a lower-return, older, existing business. The twin techniques are first to make the all-important right-hand side of the lever—the longer part to the right of the fulcrum—as long as feasible, and second to make the opportunity quickly grow as large as it can. This is done by constantly pursuing new high-margin growth opportunities and, when necessary, abandoning low-margin ones.

Abandoning a low-return business to redeploy the assets to a high-margin business gives you a double leverage effect. First, you reduce the amount of lower-margin existing business. Second, you may not need to increase your assets much at all to accommodate the growing new business.

What TechnoLeverage Means for Corporate Planning

In its essence, strategy encompasses knowing who you are, where you want to go, and how you are going to get there. All the classic questions in corporate strategic planning—What is our business? Who is the customer? Who are our competitors? What are they up to? What is our goal for next year? Where do we want to be in five years?—are geared to helping management frame and decide these three issues: Who are we? Where are we going? How will we get there? Too often, these issues are settled by simply accepting what is available.

To start putting technology strategy into practice, ask two key questions: "What's possible?" and "What's desirable?" In other words, management must ask: "What *can* we do?" and "What *should* we do?" It takes concentrated thinking to develop serious answers to these questions. To answer them in gut reaction or without good alternatives is to abdicate a large chunk of managerial responsibility.

The last two questions should be asked in the order given. The first question, What's possible?, enlarges the field. It lets you look at your capabilities and choices broadly. However, there is also a narrowing function within this question, in the sense that you are also asking, "What can we do with what we've got?" Although brainstorming, expert polling, benchmarking, best industry practices, and other creative, open-ended approaches should play a role in answering this question, the possibilities must be approached realistically. Perhaps the question should be phrased, "What's possible *given the available capabilities*?" so that the answers yield useful insights. As we see in later chapters, there are several useful ways of defining, locating, and accessing "the available capabilities."

You are not asking this first question so as to find fantastic, far-afield possibilities. Yes, it is possible for Intel to enter the toothpaste business. Yes, Procter & Gamble can develop light-emitting diodes if it really wants to. But these moves would be inconsistent with their core capabilities and how they add value. The question should be framed to yield strategically useful answers. For example, it would be both possible and reasonable for Intel to design new functions such as teleconferencing onto their chips. It is merely possible (but not advisable) for them to get into manufacturing complete personal computers.

The second question, What's desirable?, translates to What should we do *given the marketplace and our capabilities?* This is to say, from the array of possible things that we could do with our capabilities and our technology, which ones make sense? Which ones does the market want? What could we actually sell? Which things can we make money doing? What should our priorities be?

The two questions work together. The first one enlarges the pool, and the second narrows it down to the most appropriate choices. If you have a set of successive hurdles, a set of sequential tests that you can apply to ideas, activities, plans, moves, applications, products, and investments, then you have the beginnings of a disciplined process. Such discipline is essential to increasing the chance of success and decreasing the chance of failure.

Make Sure R&D Includes Reality and Direction

We talk more in Chapter 9 about organizational issues, but there is a key point worth making here about running a research and development organization. R&D and its management team play an important role in TechnoLeverage. Their abilities, exposure, point of view, and depth affect the choice and implementation toward the new directions that the company considers when pursuing higher-margin opportunities out on the right-hand side of the lever.

One simple step toward successful new business and away from chancy TechnoMarvels is to create a managerial partnership between a technologist and a business development person. In such a pairing, both individuals contribute their special skills. Pairing keeps business managers in touch with technological capabilities and advances, while keeping technologists in touch with business reality and issues.

This kind of balance between technology and business development is how the real winners in technology consistently achieve extraordinary returns. Such techniques can be successfully practiced by many more companies. There are few good models of how to manage technology. The high-tech companies that consistently achieve high returns understand how technology leverages their performance, and their business practices show us how to construct this leverage.

Review and Preview

In this chapter we've seen that strong forces oppose the growth and profitability of most businesses and that technology can enable a company to overcome those forces. Yet technology itself is a spirited steed and poses both risks and rewards for managers who choose to ride it. The primary sources of risk are the novelty, rapid change, sophistication, costs, competitive alternatives, and difficulty of managing most technologies. These characteristics of technology cause the median return of companies that make and sell technology to be lower than the median return of nontechnology companies.

However, some technology companies do get it right. Some of them do consistently produce the extraordinarily high returns one would expect technology to yield. These organizations do some combination of the following:

- Concentrate on core technological skills.
- Invest consistently to enhance those skills.
- Search widely for applications and markets.
- Target high-growth, high-profit markets.
- Strive for dominance or leadership in their area.
- Sell whole products and complete systems.
- Price aggressively.
- Abandon low-profit, low-growth markets.

We've seen how technology allows them to compete in new ways along the dimension of uniqueness. Pursuing innovative and desirable uniqueness introduces our second TechnoLeverage Tip: *Use technology to create differentiation on which to build high-margin, high-growth new business.* We've also seen the mechanics and the math of how this improvement in returns affects the companies that pursue TechnoLeverage.

The next seven chapters of this book show you how to adapt these practices to your own company and carry them out, regardless of your line of business.

Next, in Chapter 3, we examine one of the most important of these practices: the search for the best applications and markets. We also study the issue of identifying and nurturing your company's core technology.

CHAPTER 3

Search Strategies

Finding the Best Applications

At the heart of an inspired technology strategy lies relentless pursuit of the best applications for a company's technology. Putting TechnoLeverage to work depends upon regularly finding applications where your technologies solve problems, add value, produce distinctive products, win customers, and generate high profits for an extended period of time. A TechnoLeverage *search strategy* is a systematic method of identifying these attractive applications and the markets associated with them. Searching for new applications represents the most important component of a technology strategy. In this chapter, we explore how to create and implement a productive search strategy.

Why Search for Applications?

The perpetual need for new applications is dictated by the downward slope of gross margin in the technology applications spectrum (Figure 1.1). A company using technology is like a youngster on a sled, who enjoys a ride down the hill and then must walk back up the hill to repeat the ride. The walk up is essential to the ride down. Having achieved wide acceptance, any application for technology declines in profitability as the gross margin inevitably dwindles. When that occurs, too many companies fall into the trap of investing more money to reduce their per-unit costs and maintain profitability in the face of lower and lower returns. Often they do this because they believe they have no other options. Searching for new applications provides a range of options, alternatives to the low-margin, low-value-added business environment the company faces in its immediate future. The more choices you have, the more opportunities you have to grow your bottom line.

A search for applications has three goals:

1. Identify opportunities to sell new products or improved variations of existing products.
2. Find higher-margin ways to replace existing products.
3. Compare opportunities.

You *improve your existing products* by searching for ways to enhance their performance characteristics—make them faster, stronger, larger, smaller, smoother, more flexible, more durable—and for ways to extend the applications themselves into related functions. Many customers require better-than-average performance, and some will pay handsomely for it. Improved performance adds considerable value if done properly and as customers desire.

Extending applications to adjacent functions is another excellent search strategy, as we saw in the case of Husky Injection Molding in Chapter 1. By moving outward to related activities, guided by its customers' adjacent processes and system-performance needs, Husky wound up in five different but closely related businesses. All of these moves added value for the customer, with each move exploiting and expanding Husky's core technology. Intel does the same thing differently by designing more functions onto the chip itself, as it spots other activities such as multimedia that computer users would like to perform.

You *replace your existing products* by discovering how to take your technology in new directions and into new markets. The experience of Otis Specialty Papers in Jay, Maine, now a division of Wausau-Mosinee Paper, demonstrates this search approach.

Otis made a product called release base, the backing for self-stick labels. The profit margins in the label business were OK but not likely to improve because more capacity kept coming into the market. Otis's capacity was constrained so it couldn't just produce more to expand its profits. The lowest-profit item Otis made was a standard-grade writing-tablet paper, which produced (forgive me) paper-thin margins. Otis believed it could enrich its margins by dropping the tablet paper and replacing it with a more profitable product. In the paper business, this is called "sweetening the mix."

Management asked, "What else can we do with what we have?" After a search for suitable and more profitable applications, they decided to manufacture coated ink-jet paper. Otis had a coating process it could apply to ink-jet paper. Over a period of thirteen months, they developed a product for this market—at *triple* the margins of their base business. In our leverage model, this is a clear case of extending the length of the lever to the right and moving the fulcrum point to the left. Despite its small size relative to the whole company, the ink-jet paper product line significantly improved overall financial performance.

You compare your existing opportunities to new ones by developing alternatives and then analyzing and ranking them systematically. This evaluation involves finding ways to take your technology—existing, new, bought, or borrowed—in new directions and into new markets.

Here is an example that we return to later: A division of a large European company had a ticket-vending technology, embodied in a piece of equipment and information systems components that generated bus and railroad tickets. The equipment tracked seating plans; communicated with the home office; diagnosed itself when it malfunctioned; took credit cards, bills, and coins; made change; figured how much cash it contained and called the main office when it needed to be emptied; offered special discounts; and was completely secure from vandalism and theft.

The company had sold these systems only to intercity bus lines and commuter railroads; the total market of some fifty customers was depressingly small. The question was, Where else could they apply this technology? The answer was, surprisingly enough, in the entertainment industry.

A broad array of amusement, leisure, and recreation organizations were looking for new solutions for on-site, remote, and advance ticket sales: movies, theaters, concert halls, theme parks, ski lifts, and sporting events. Convenience retailers wanted to use the equipment to sell scratch-and-win lottery tickets. All these businesses were potentially interested in the unusual value that this company brought to them—secure, low-labor-content retailing of tickets. The expansion opportunity was both bigger and more profitable than the company's base business! Which raises our third TechnoLeverage tip:

> *TIP NUMBER THREE:* **Thorough technology application searches produce pleasant surprises.**

This is why you should take them in all possible directions.

The best search strategy accommodates all three main goals—improvement, replacement, and comparison—although accomplishing just one of them is enough.

Impatience Is a Virtue

Managerial impatience with the status quo is usually a key motivator in application searches. A company need not face imminent ruin to embark on a search for new applications, but management does have to want some things to change. Think about what ignites a search, any search: You want to find something, either because you've lost it (such as a portion of your market or profitability) or because you want something you don't have (such as a new market or more profitability). The intensity of your search is usually proportionate to the intensity of your desire for what you seek.

Managerial impatience and intense goal orientation distinguish an inspired search strategy from a mundane effort at expansion. A company may believe that it has a "search strategy" because it monitors competitors' new offerings or employs a director of product development. In isolation, such search efforts are usually defensive, often perfunctory, and routinely underfunded. The same usually applies to management's response to those efforts. Unless top management is completely committed to new markets, nothing new will appear.

A search strategy must seize the initiative. It must seek something, find something, create something. If an applications search strategy has a mission, it energizes itself and the organization around it. Without the spark of top-level impatience, restlessness, and goal orientation, searches for new applications become a routine, go-home-at-4:30 endeavor. Top-management dissatisfaction ignites the search.

Seizing the initiative with a search means going beyond typical product development tools such as product-line extensions. Much of the product proliferation that occurs in consumer packaged goods, such as food and personal care items, is of this type. It usually results in trivial permutations: Ranch Flavor, Cool Ranch, Spicy Cool Ranch, Lowfat Spicy Cool Ranch, Lowfat Spicy Cool Ranch with Avocado, and so on. The distinction between product-line extensions and a true application search is worth drawing. Much of what passes for product development in consumer and industrial markets is reactive and marginal. It winds up merely moving money and resources around. Instead, try making a concerted effort to find another problem to solve. This way, you stand a better chance of creating new value and leveraging your assets significantly. I like Cool Ranch myself, but here we're talking about something more profound.

What Do You Have That Works?

As noted, a technology company's core technologies, the prime source of its value-added, find expression in applications by solving customers' problems. It is rare for a company not to have problems to work on. The world is full of unsolved problems. If you look, you will find them. Of course, the trick for TechnoLeverage is to solve problems and make money at the same time.

An ideal technology to work with has wide applicability. An applications-rich technology seems readily capable of solving a large range of problems. Think of how many ways duct tape works as a repair tool; think of Velcro as a fastener. Think of the incredible number of uses for electronic circuitry on a silicon chip: computers, televisions, networking, automotive electronicsx communications, instruments, power supplies, graphics. The silicon chip, like duct tape, has the characteristic of "going in all directions," as I call it. It gets pressed into service everywhere.

In sharp contrast, consider a company in the southwestern United States whose core technology is instrumentation that can distinguish among electrical signals picoseconds apart. (A picosec-

Figure 3.1: Applications "Tree" Diagram

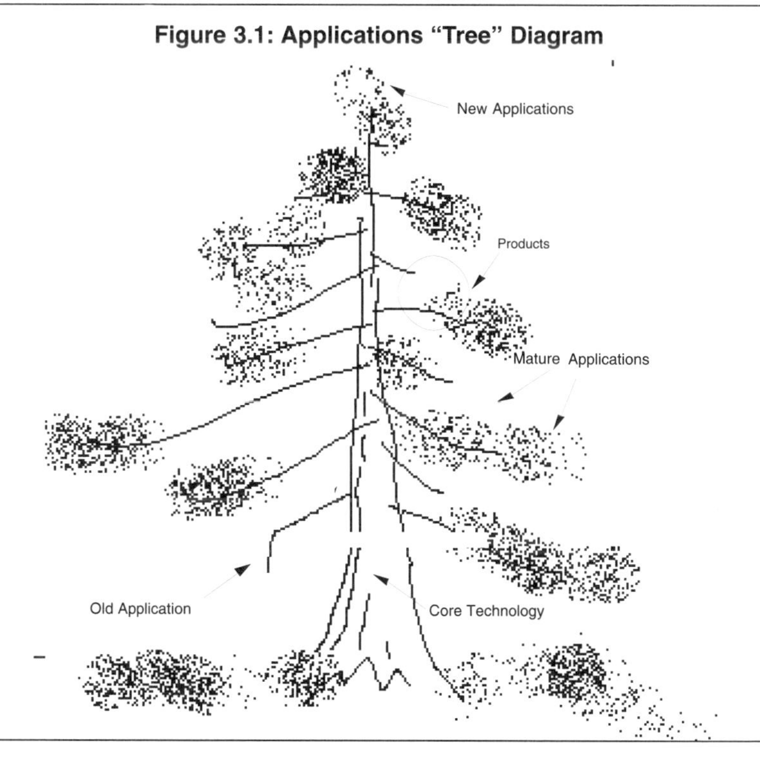

ond is one-trillionth of a second.) This technology first found application in measuring the formation of the shock wave from atomic bomb blasts. It's a very accurate measurement capability, very sensitive. But people generally have few problems that require the ability to measure trillionths of a second. So the technology likely faces a longer than normal search for its next application.

Think of the core technology of a company as the trunk of a tree, from whose main stem everything else grows, as in Figure 3.1. The core technology is the central scientific or engineering knowledge upon which the company's long-term success depends. Applications of a technology grow, mature, and die, paralleling the lives of the branches. Products have a still shorter life cycle. Core technologies, however, can prosper for decades, particularly if they are rich in applications. Applications and products are manifestations of core technologies, but they are distinct from them.

A search strategy finds applications of the technology that can sustain the company's growth. Since sustaining the company's growth must be among management's major responsibilities, a search strategy must command top-management attention. Here are two contrasting corporate examples.

In fifty years, Polaroid has developed few applications of its core technologies, which cover design and manufacture of instant color imaging (to its credit, at least the company tried with Polavision, as described in Chapter 2). Those skills made Polaroid the world leader in a narrow field called mordant chemistry. Mordants make colors stable by combining with dyes and making them chemically inert. Mordants make ink-jet ink permanent. But is Polaroid anywhere to be found in the ink-jet world? No. Polaroid lacks an effective search strategy that would have found this new application for its skills.

In contrast, consider Fiber Spar and Tube Corp., known as Fiberspar, an advanced-composite manufacturer based on Cape Cod. Fiberspar braids high-strength fibers and coats them with epoxy to produce strong, ultralight products. Fiberspar started out making windsurfing masts, the original application. Not wanting to be limited to the fickle sporting goods market, Fiberspar has repeatedly explored new applications. Many of these have not panned out and were jettisoned, either after initial market research or during trial production and marketing. Fiberspar's search examined golf club shafts (no profit), arrow shafts (tiny market), idler rollers for the paper and film-handling industries (commodity-pricing mentality). These were nice technical ideas, but they made poor business prospects. Still, Fiberspar persisted, eventually finding two new applications: hockey sticks (in production, but the jury is still out), and composite coiled tubing (big technical and business successes, under way). The composite coiled tubing division, which replaces steel tubing in down-hole oil service, is already half as big as the rest of the company in terms of sales after one year of production. Fiberspar kept searching for new applications to find something really good. And it surely won't stop at coiled tubing.

As you set up a search strategy, it's crucial to concentrate on applying the technological skills that add the most value, including others' technologies. Intel is very good at applying the technologies of others. The company's core technologies, its main source of value-added, consist of designing and making microprocessor chips. Intel engineers the architecture and essential logic of the chip and fabricates the chip itself. But to this capability Intel applies a staggering array of other companies' technologies. To Intel, these are *supporting technologies.*

Intel's supporting technologies encompass purity, cleanliness, speed, precision, miniaturization, and an array of complex, multifunction mathematical processes. These take many forms: extremely precise saws for cutting large silicon wafers; machinery for polishing; special tables for positioning the wafers properly; chemicals for processing pure to one part per quadrillion; atmospheric control systems that perfectly maintain air pressure, velocity, and temperature; math routines for laying out the chip circuitry; and testing equipment for quality control. Most are bought from the outside.

Intel is a great fabricator of chips, but it's also a great assembler of others' sophisticated manufacturing know-how. Intel designs and fabricates the chip; this is how they add value. But they outsource the development of everything else. That's smart, and it clarifies the distinction between core and supporting technologies. Core technologies are the most valuable ones you own, the ones you seek to protect. Supporting technologies are those you use but need not own. Businesses should capitalize on the best available technologies regardless of who owns them.

Companies with a clear understanding of their technologies and how best to apply them succeed. They sustain their success by repeatedly finding new applications that reinforce their chosen position in the marketplace. In contrast, companies that cannot figure out how to apply more widely whatever they do stagnate.

Choosing and Maintaining a Position in a Zone

The search for applications must focus on one of the target zones that compose the technology applications spectrum in Figure 3.2: uniques, exotics, specialties, and commodities.

Let's define these zones more precisely.

- *Uniques* are brand-new offerings. Typically, these are the first items produced in their category, and so they define a new need. For a time, they represent minimonopolies. The producer sells to end users who value the product or

Figure 3.2: Picking Your Zone on the Technology Applications Spectrum

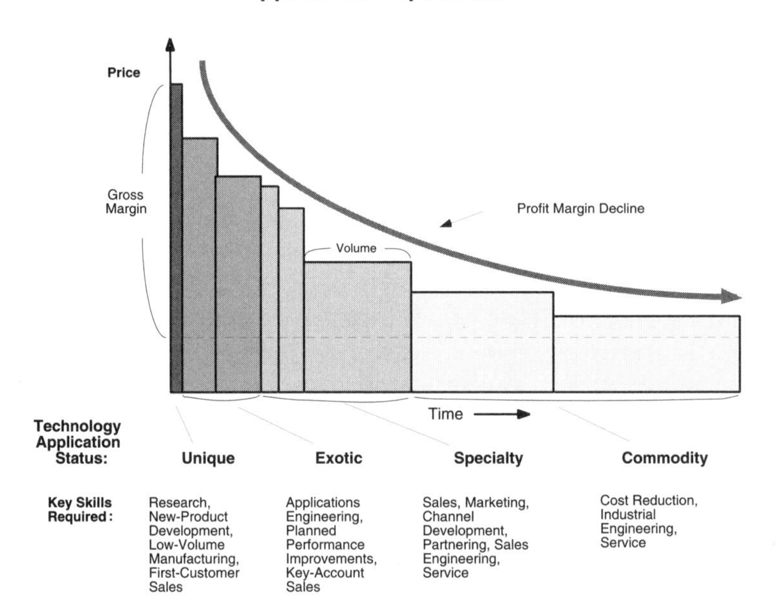

service highly. Thus margins on uniques are often sky-high, and volumes for these unfamiliar solutions are usually low while the products gain acceptance. Offering uniques appeals to companies employing very original thinkers and having lots of inventive power. When an exceptional company like Intel offers something unique like the Pentium chip and then produces millions of them, money cascades into the business.

- *Exotics* are highly differentiated, high-value-added products that are still new. The market for exotics is narrow but has begun to acknowledge great need for the product, so new producers have joined the leader. As a result, capacity expands, and prices and margins begin declining. A good example is a biotechnology-produced agricultural product, say herbicide-resistant corn and soybeans. As specialized as these exotic applications are, they appeal to a broader market than do products in the unique category. Advanced engineering skills are required to make major modifications to the product so that it is usable by more customers. Exotics are no longer a "onesies and twosies" market. Market acceptance and growth are under way.

- *Specialties* are more familiar applications of a technology. In the specialty zone these are usually still niche products, still somewhat differentiated, still relatively high-value-added. They enjoy a broader market than exotics do. Customers may sense that this product is getting more price-competitive, feel more confident about using it, and sniff around more energetically for comparison price quotes. More producers have figured out how to make it, and the real profits go to sales-driven organizations. Specialty prices and margins are well below those of exotics and above commodities. Laminated and coated paper products, such as Avery Dennison's ink-jet labels, are specialties. Avery can sell thousands of tons of its labels yet still achieve above-average margins.

- *Commodities* are products whose use and performance are fundamentally uniform across the range of suppliers. They are undifferentiated, and they command lower gross margins than do specialties. Competition among producers usually centers on price, and perhaps on service and relationships. Engineered raw materials, particularly petrochemicals, fit into this category. Other examples are such familiar industrial products as fractional horsepower motors, LAN cabling, and agricultural supplies.

The technology applications spectrum reproduced in Figure 3.2 illustrates the dynamic forces that affect a successful product throughout its life. Market forces pull successful products rightward across the spectrum, toward the next category. As a company follows its products, margins decline. Sometimes volume growth compensates for the lower margins, sometimes not. Companies employ a search strategy to resist the forces pulling them toward lower margins. If we think of the gross margin curve, we see that companies resist the margin decline by working back uphill, finding higher-value-added things to do. A company can make money anywhere along this spectrum, even in commodity products. However, technology strategy and corporate culture demand that management target a position on the spectrum—within the unique, exotic, specialty, or commodity zones—and then apply their technologies to maintaining that position.

To maintain a position in dynamic markets, one must balance out the forces of invigoration and decline. It feels like walking up a down escalator; you must keep walking just to stay in place. A company then achieves dominance by excelling at activities consistent with its chosen position on the application spectrum.

Don't Try to Play in More Than Two Zones

It is nearly impossible for a successful corporate culture to span more than two of these zones because they require such different strategies, actions, and mind-sets. Ignoring this fact creates numerous problems. Management teams overestimate their ability to change their companies, their culture, most of their personnel, and virtually all of their operating methods. But such changes are in fact required for success in the adjacent zone.

The wisest corporate strategy for most companies is to resist movement rightward across the application spectrum. This means that exotic and specialty businesses must repeatedly find new and higher-value applications. They must also resist commoditization of their current products. Commodity-focused companies should try to accelerate commoditization of what are now exotics and specialties, standardizing product features and lowering costs and prices. Commodity producers can use technology to reduce production costs, and to add value to service or quality.

How Does the Market Affect Application Searches?

When you consider a new application for your technology, you need to decide where to start. Understanding potential buyer behavior helps greatly. Let's look at two familiar analytical tools: the adoption-diffusion model and the product life-cycle curve. They've been around for years and help explain market reactions to new applications of technology.

The adoption-diffusion model was developed by the U.S. Department of Agriculture in the 1930s to explain why farmers were reluctant to adopt new technologies such as hybrid crops and husbandry techniques such as contour plowing. It depicts how groups of buyers accustom themselves to new applications of technology as an application moves from novelty toward full market acceptance. Individuals and businesses adopt new technologies in a predictable pattern that follows a normal distribution, giving it the appearance of a bell-shaped curve (Figure 3.3). The adoption-diffusion model divides users into five classes, according to how they sequentially consider and begin using a new technology.

The adoption-diffusion model demonstrates how innovators, the roughly 3 percent of buyers with a need for one-of-a-kind products, always buy into a new technology first. Then the early adopters, who also desire novelty but are more hesitant than innovators, purchase it next. As the new application of technology

Figure 3.3: Adoption-Diffusion Model, Characterizing Technology Adopters

The Adoption-Diffusion Curve

Technology Adoption Behavior Groups

Innovators	Early Adopters	Early Majority	Late Majority	Laggards
Want Highest Performance Available	Want to Be Leaders In Their Business	Want to Stay Ahead, but Not Be the Guinea Pig	Don't Want to Fall Too Far Behind	Reluctant to Change
Most Independent	Want to See It Done Before They Buy It	Want to Buy From Established, Accessible Suppliers	Want No Glitches; Need a Thoroughly Proven Performer	Most Fearful
Highly Knowledgeable About Their Their Business	Among the More Profitable	Product Must Be Easy to Buy	Most Price-Conscious Buyers of All	Least Well Informed About Their Business
Best Informed, Most Profitable	Interested in New Solutions	Want to Know You Have Sold Lots of Them, or Have Done It Often	Want to Buy Locally	Adopt Innovations When They Can No Longer Be Avoided
Eager for New Solutions	Very Closely Watched by the Early Majority		Absolute Simplicity of Purchase and Use is Critical	Often the Least Profitable
Will Buy an Unproven Concept				

proves itself useful and is more widely understood, the early majority adopt it. They are followed by the late majority, which buys into the new technology only after it has become a standard. The laggards buy only after it has become completely unavoidable, a widely accepted item, and no longer new.

Each watches the group to its left, with the innovators thinking for themselves when it comes to a new application. Since the product has no track record to judge, innovators are influenced by reviews, the seller's reputation, their own insights, and expert salespeople's careful logic about achieving results. Innovators had car phones twenty years ago, when they cost $3,500. They bought them despite the price because they understood how to make money with them.

Early adopters watch innovators, and the early majority watches early adopters. By the time the early majority is fully on board, the product is on its way to a large market. Volume rises, and prices and margins fall. The late majority adopts the innovation. Finally the laggards are convinced, but often because adoption is inevitable. Laggards are buying their first color TVs nowadays because it's getting difficult to find a really good black-and-white set.

This adoption process is synchronized with the well-known product life cycle, shown in Figure 3.4. The product life-cycle curve depicts the stages of a product's life:

- Introduction: adoption by innovators

- Growth: acceptance by early adopters and some of the early majority

Figure 3.4: The Product Life-Cycle Curve

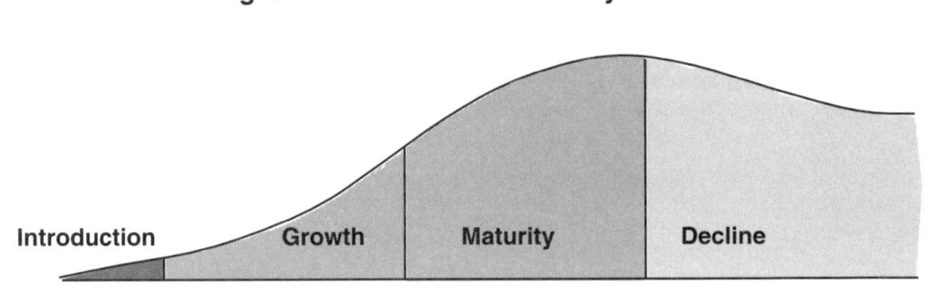

- Maturity: adoption by the rest of the early majority and part of the late majority

- Decline: acceptance by the rest of the late majority and the laggards

Once the laggards have adopted the product, the market is completely mature because all prospective buyers have entered the market. How different from the exciting days when the product and the application were new!

The product life cycle depicts the path to sales maturity and decline, but it doesn't show the pressure on prices and margins. The TAS plays that role.

Together, these three curves provide a simple and accurate qualitative forecasting tool. They give you a useful way to think about your applications, product development, markets, plants, finances, staff, and operations:

- The product life cycle helps you chart your product's growth and maturation.

- The adoption-diffusion model helps you understand your customers at each stage of market development.

- The technology applications spectrum addresses financial issues such as margin decline and required capital investment, thus providing the missing link between the other two.

Thus the TAS, which we developed from our work with clients, connects the market's behavior (the adoption-diffusion model) with the product's behavior (the life cycle). You can think of the three together in this way (Figure 3.5).

In Figure 3.5 all three planning curves have been aligned and two dotted lines drawn over them. One dotted line shows the convergence in the case of fiber-optic cable used for cable television, and the other shows fiber-optic cable used for telephone. Follow the lines to the three curves to anticipate the dif-

Figure 3.5: The Three Planning Curves Aligned on Two Sample Technologies

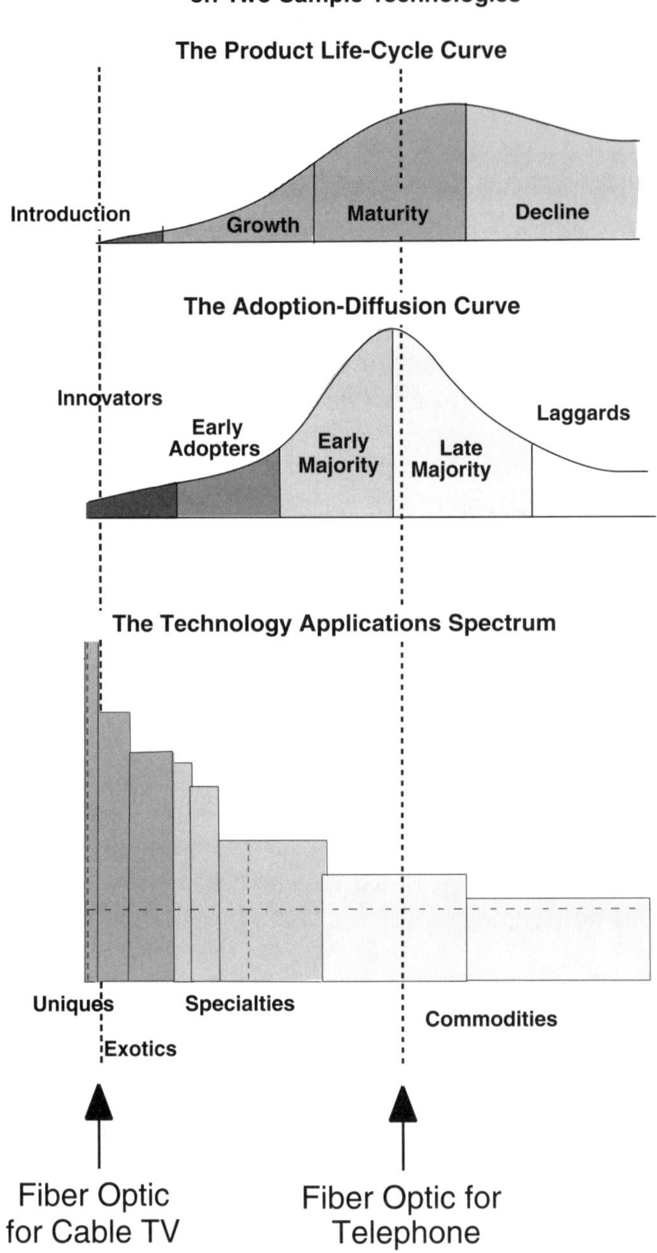

ferent reactions, responses, market conditions, and financial prospects a supplier might face in the case of these sample applications.

Nuts and Bolts of an Application Search

A company must choose among the unique, exotic, specialty, and commodity zones on the TAS to find a zone that is consistent with its culture, capabilities, and technology. Then management must devise a robust process to find new work there and offset the forces pulling its margins downhill. An effective application search strategy does just that.

The application search should have three components:

1. Scanning
2. Elaboration
3. Balanced evaluation

Let's examine each of these in turn.

Scanning: What's Going On Out There?

The purpose of scanning is to search the horizon for developments that can affect you or that you can capitalize upon. Good scanning—a formal, wide-angle, ongoing information gathering process—is surprisingly rare in business. This is too bad, because scanning is inexpensive and yields benefits far in excess of its costs. The benefits come in the form of opportunities identified as well as mistakes avoided. Scanning the horizon would have helped Polaroid avoid its painful $50 million Polavision episode.

If a company is to acquire this information steadily, someone must be responsible for scanning and have a process to use. In practice, most companies gather information willy-nilly. Willy-nilly comes in three flavors:

1. *The "strategic treasure hunt."* This sporadic effort waxes annually in anticipation of the strategic planning process and then wanes. Someplace on the planning template

one must cover the issue of "what's happening," so some poor soul rushes to find out. Information gathered for the annual strategic plan is often fragmentary and seldom reveals the underlying patterns.

2. *The "snoop group."* Sometimes scanning is relegated to a competitive assessment function, whose output is frequently ignored by senior management. Competitive assessment tends to focus only on the current direct competitors. These competitors should be monitored, of course, but as we see in Chapter 4, the kind of competition that technology produces renders routine monitoring of known competitors insufficient.

3. *The "self-appointed hobbyist."* In other cases an individual carries out the scanning function for the company, sometimes out of curiosity, sometimes out of exasperation. Such entrepreneurial searches, while often inspired, are always part-time and usually random and individualized. If this person leaves, the function promptly withers.

Scanning must be done broadly, regularly, and permanently, just as the watch is kept on a naval warship, for two reasons. First, you don't want to miss anything important. Second, you want to encourage making connections between isolated observations, new concepts, unexpected developments, unfolding processes, and poorly expressed needs. This calls for "loading the hopper" with a lot of varied raw material. Scanning consistently and broadly helps you do that.

A Solution

An excellent place to lodge a scanning effort is the new-business development group, especially if the group reports high up in the organization. Energetic newcomers with some seasoning and a well-thought-out checklist to pursue usually do the best job and have the advantage of an open mind. Scanning is a good training ground, and it gives a high-potential new employee useful exposure to senior management.

One way to structure this scanning process is to locate the "promontories" for your discipline. Promontories are vantage points from which you can see a varied stream of ideas, information, innovations, people, events, developments, and disciplines. Promontories may be publications, organizations, committee memberships, the contributions of certain experts, databases—even professional gatherings and conferences.

One very useful promontory is *Science News*, a weekly publication that reports on advances in all the sciences. It provides snapshots of recent discoveries in an amazing range of scientific disciplines. Most articles are short, half a page or less. They are particularly strong in the areas of materials and materials development, covering everything from ceramics, metals, and submicron photolithography to blue-green algae, earthquake damage, and moon rock fracture patterns. Their Website at www.sciencenews.org is very useful.

Popular Science is completely different. *Pop Sci* reports on numerous areas where various technologies have been commercialized and popularized—tools, vehicles, instruments, home improvement, home entertainment—adding up to a rich and unpredictable mix. Serendipity certainly reads *Popular Science*.

If you like your scanning heavy and in black and white, reading patent filings and trademarks can yield enormously valuable technological and competitive information. This kind of information is available in the *Patent, Trademark, and Copyright Journal*, published weekly by the Bureau of National Affairs. You will find a database of trademarks in major libraries in hard copy, on CD-ROM, or on-line. Another promontory is *Companies and Their Brands*, published annually by Gale Research.

The Internet is growing in breadth and importance so fast that it seems to double in value monthly. A powerful and useful tool TMG has found is a super search engine, www.dogpile.com. Enter the key search terms you are looking for into dogpile.com (there must be a more mellifluous name available) and it immediately searches a number of other search engines for the reference you seek, important ones including Lycos, Yahoo!, Thunderstone, and AltaVista. For the technology application scanner, the attractive feature of this approach is that each of the "lower-level" search engines has a particular way of searching for key words in Web pages. Each has its strengths while missing certain kinds of references. By using a number of search engines, Dogpile gives you a much better chance of finding exactly what you want on the Internet.

Believe it or not, annual reports definitely warrant scanning. Although most annual reports deliver little insight, in my experience about 10 percent of them have exceptional educational value. They tell you what other management teams think is possible. Once you've identified the 10 percent in the universe of public companies that are relevant to you, scanning their annual reports should repay the time invested.

Conferences also represent promontories. They bring together people and ideas from all over the world. You can't attend them all, but the awareness and contacts that you gain from a copy of the conference proceedings often outweighs the small cost. Divide them up among your staff.

Finally, no scanning effort worthy of the name can omit competitors, customers, prospects, and suppliers. The press is a good source in these areas, and the sales force can be an even better one. Too many organizations ignore clear messages from their sales force. Although the information can be inaccurate, it's a good idea to carefully consider observations from the field.

Elaboration: Make Something of It

Now that you have the information you've sought through scanning, what should you do with it? Here's a technique to help you to build on the information you've gathered.

Elaboration is the act of breaking a big thing into its component parts and forcing your organization to see all the possible elements. In business searches, elaboration is a nearly flawless tool for finding the overlooked nugget of gold, the great opportunity. It's a way to take the results of your application search as possibilities and then finding and filling in the gaps to generate all the possible applications of your core technologies. Elaboration helps you build up the detail and create a strong foundation for expansion scenarios.

The first step in elaboration is to ask what's missing. Press yourself to identify what you didn't see, especially what you expected to find and didn't locate. Take care to identify missing elements so that you can confirm their absence or discover why they are missing.

Another method of elaboration I call "chaining," in which a company moves out from existing applications to other functions. This is what Husky Injection Molding did, one step at a time. A chain is built by forging links from existing applications to those nearby. Recall that in Husky's chaining the company examined what its customers were doing and thus realized that hot runners and product-handling equipment fell short of the performance of Husky's molds. The shortfall presented opportunities.

In elaboration, consider your technology or application from the outside, as a stranger might. Put aside your intimate knowledge of the product or service and ask:

1. Who else could possibly use this?
2. What possible use could I make of this if I were a farmer, an executive, a banker, a pilot, a homemaker, a doctor, a librarian, an accountant, a publisher, a transportation manager, a shipper, a contractor?
3. How would I customize it?
4. What current applications in other areas could this replace?
5. Which applications of ours or someone else's could this be combined with to create something new and useful?
6. How could this be used to reduce costs in some area?
7. What would happen if we made this faster, stronger, lighter, cleaner, clearer, bigger, smaller, more durable, more flexible?
8. How large is the problem that this application would address?
9. How costly is the problem for those who have it?
10. How widespread is the problem? Is it local, regional, national, international, or global?

11 Does the problem affect one group of consumers or one industry, or does it cut across groups? If it affects only one group, is it a large group?

12 How easy would it be to copy this application? How could we protect ourselves if we had developed it?

Then summarize by sketching the answer to the question, "What problems can I really solve with this?"

The purpose of elaboration is to challenge yourself and your organization to find really new opportunities by making unfamiliar connections. If you are dealing with a technology that is even moderately application-rich and you do a good job of scanning, you should be able to elaborate potentially useful applications. This can be done even with well-established technologies such as rechargeable batteries (Black & Decker does an excellent job with them, solving all kinds of common problems with cordless screwdrivers, Dustbuster vacuums, drills, hedge trimmers, and so on). Your goal in elaboration is to drive detail into your set of opportunities, detail that is not there now.

Balanced Evaluation: Over the Hurdles

You have to test your new opportunities, and the best way is balanced evaluation. This is no time for simplistic return-on-investment rules, rigid hurdle rates, or other one-number fallacies. Balanced evaluation takes into account the fact that business is complex and your knowledge is incomplete. Recall that one goal of an application search is to compare. Now we have to compare potential applications to existing ones and to one another. A balanced evaluation does this.

One thing is certain: The overall new-product failure rate (reportedly about 90 percent) would be much lower, and the return on assets of technology companies much higher, if these criteria were applied consistently to each potential application:

- What is the demonstrated level of customer interest in this application?

- How large is the true potential market?

- What are our chances of securing a leadership position for some period?
- How long will it take us to bring a product offering to market?
- Can we engineer the application ourselves? If not, how can we get cost-effective help?
- Are we set up to sell something in this application now? If not, how can we obtain cost-effective distribution?
- How much does this application leverage our technology?

Here are some ways to develop useful answers to these questions:

- Gauge customer interest.
- Size up markets by using a bottom-up approach.
- Estimate time-to-market carefully.
- Calculate engineering and sales requirements realistically.
- Assess competitors thoroughly.

In *gauging customer interest*, you need external validation. The quickest way to get this validation is with a concept test, whereby you describe the application or product in detail to a potential user (usually an innovator or early adopter). If at all possible, show a prototype; use a confidentiality agreement if you have to. If innovators or early adopters can't understand it or can't see an important place for it in their operations, you probably don't have a useful application. (However, keep Citicorp's ATM experience in mind. In pursuing TechnoLeverage, you must often lead the customer.)

Sizing markets represents the trickiest and riskiest part of application assessment. This stage of the process magnifies your natural optimism. You want to see a large market; you want to believe that everyone who has the problem would buy your application. (In fact, as Bill Gates and Andy Grove can tell you, that does happen once in a while.) But as difficult as it may be, you must be realistic about potential market size.

Overestimates of market size usually stem from top-down methods of market estimation. Top-down calculations usually start with something like the U.S. economy or the number of households in the country and then apply some arbitrary percentage to that base. One then assumes that some percentage of them will become customers. This really amounts to guessing, without understanding the underlying decisions being made.

A bottom-up approach generally yields better estimates of market size. A bottom-up approach roughly simulates decisions made by individual potential users. You outline the decision process the user or the buyer goes through. Do you have a new card stock that prints glow-in-the-dark safety labels for factories, on a regular ink-jet printer? OK, what proportion of the surveyed safety department decision makers say they'd like to see them used in their shops? How many labels would they make per year? Have they seen the results? Will they rebuy? In what proportions? Will more safety departments try them after they see them written up in the magazines? Will catalogs carry them? Is the hospitality industry interested? What proportion of hotels say they will use them? What volume do they think they'll consume? A bottom-up approach arranges these estimates into a logical process that mirrors actual experience. Then you can use top-down calculations as a double-check.

We examine technological leadership—its rewards, costs, and risks—in Chapter 5. Despite the challenges, most businesses should try for leadership. The rewards of leadership outweigh the risks and costs for a well-run business with a new, high-value-added application.

Time-to-market is also tricky to estimate. Think of all the technology products that fall behind their initial release dates. There is art as well as science in bringing a product to market: the art of managing program teams toward ambitious goals, and the science of building something that works. We talk more about work groups in Chapter 9. Current best practices in this area center on small cross-functional teams, and this kind of team should be the source of time-to-market estimates.

Engineering requirements and *sales requirements* must be based upon realistic assessments of company capabilities. These can be relatively easy to assess because you know your engineering staff and their capabilities. You know whether or not you have a distribution channel to hospitals or retailers or whichever other market you may be targeting. The most common pitfall is underestimating the cost and effort required to close gaps in engineering and sales resources. Sometimes it takes years to find the right individuals in sales and engineering.

Assessing competitors can be tough, even with all the great new techniques and the information they generate. We cover ways of assessing competitors in Chapter 4. The best starting point is good scanning.

In Chapter 6, we examine various techniques for *leveraging technology.* But when assessing a potential application, think in terms of judging the "lift" in earnings that the product brings. Think also of how you can achieve leverage down the road, for example, through licensing, partnering, and other tactics (covered in Chapter 6).

What About Traditional Investment Analysis?

Traditional investment analysis tools, such as internal rate of return and payback period, tend not to work very well in situations involving new technological applications. These tools should not necessarily be avoided. They are great for established industries such as real estate and finance. But their limitations become clearer in new situations. Here's an example of how traditional financial analysis falls short when technology is the subject.

Back in the early 1990s, Rubbermaid used payback analysis to consider a significant purchase of injection molding machines. Rubbermaid's internal analyses showed that the investment would not pay back within its stated corporate time constraint. But the conventional payback analytic approach did not account for several key facts.

First, Wal-Mart, a major Rubbermaid customer, was establishing just-in-time (JIT) inventory practices. Wal-Mart required Rubbermaid to ship more products in smaller quantities more frequently. Stated in the extreme, if a store sold a yellow dish tub, Wal-Mart wanted Rubbermaid to ship the store another yellow dish tub.

Second, Rubbermaid had explicitly stated as a corporate goal that they intended to aggressively proliferate their products in function, size, color, and other descriptors. No one at Rubbermaid had looked at the inventory and manufacturing cost of making, stocking, and shipping more variety via immediate shipments. And no one at Rubbermaid considered the conflict between limiting inventory investment in an expanding product line on the one hand and, on the other, the cost-reduction opportunities that stem from long manufacturing runs.

Third, payback analysis couldn't account for the underlying strategic issue: The molding machinery in question was the sole means to move to rapid-response manufacturing. With Wal-Mart moving to JIT practices, Rubbermaid would be hard pressed to buy enough land to inventory enough of the products it intended to proliferate so that it could ship anything Wal-Mart wanted whenever they wanted it. Rubbermaid had to disinvest in inventory and invest in flexible, fast-turnaround, short-run production machinery linked to its electronic data interchange system with Wal-Mart. This solution was not possible with the existing machinery, and payback analysis was an inappropriate test because it couldn't factor in all the strategic variables.

Simulation analysis, a form of bottom-up analysis, often does a better job in these situations, and it did so in this case. Once Rubbermaid accepted the results of this bottom-up approach as the proper analytic, they could see into the math of the problem. They went ahead and bought $35 million worth of injection molding and materials-handling equipment. With their new flexible manufacturing capability, they proceeded to proliferate product, reduce their inventory levels, and satisfy Wal-Mart's JIT requirement.

Summarizing the Results of a Balanced Evaluation

Earlier in this chapter, we discussed the European company with the ticket-vending technology. The matrix in Figure 3.6 sums up the results of an application search conducted for that company, including the summary results of the balanced evaluation mentioned above.

This kind of matrix is an excellent tool for management discussions of alternative opportunities. However, it has real value only if the underlying information is the result of disciplined effort.

Figure 3.6: Summary of Application Search for Ticket-Vending Technology

Column #	1	2	3	4	5	6	7	8	9
Criteria ► / Markets	Application	Interest	Market Size*	ABC Co. Potential*	Time Horizon	Engineering Requirements	Sales Requirements	Key Competitors**	Value of Leverage Opportunity
Gambling	Instant Lottery Tickets	Very High	$240MM	5% Technology Leadership Opportunity	Now	Low	Low	ERI	High
Movie	Admission Tickets	High	$10MM	5-15%	Now	Moderate	Moderate	Racer, Fast Ticket	Moderate
Airline Tickets--A	Tickets, Boarding Passes	Very High	$40MM	10-20%	6-12 Months	High	Moderate	Trebelda, JGT	High
Ski	Lift Tickets	Moderate	$10MM	10-20%	6-12 Months	Moderate	Low to Moderate	Fast Ticket	Moderate
Theme Park	Admission Tickets	Low to Moderate	$80MM	15-25% Technology Leadership Opportunity	12-24 Months	Moderate	Moderate	Fast Ticket	Moderate
Event	Admission Tickets	Low to Moderate	$48MM	10-20% Technology Leadership Opportunity	4-6 years	High	Moderate to High	Pass Master, PLM, Racer	High
Airline Tickets--B	Tickets, Boarding Passes	Low to Moderate	$20MM	10-20%	3-5 years	High	Low	Trebelda, JGT	High

**Estimate*
***Names disguised*

Devising and conducting an application search strategy is not easy. But it's better than living without choices, better than submitting to competitive forces, and better than watching your markets and margins erode. A vigorous search strategy for new applications is the best way to begin business renewal.

Review and Preview

This chapter focuses on a key practice in achieving TechnoLeverage: using search strategies to uncover and develop opportunities. An application search can result in improvements or replacements for existing products, and it can develop opportunities for comparison to your current options.

A company needs to choose for itself an appropriate application zone on the TAS: unique, exotic, specialty, or commodity. By beginning with a good search strategy, it can maintain that position by continually finding new applications within the zone. By repeatedly searching out and developing new applications, it avoids watching the profitability seep out of its products as competitors swarm.

The three key tools in a search strategy are scanning, elaboration, and balanced evaluation, all of which we examined here. We identified the third TechnoLeverage Tip: *Thorough technology application searches produce pleasant surprises*. That's why you should take each search in all possible directions.

In the next chapter, we look at the special kind of competition that technology and the search for applications generates. When companies take their technology in all directions, the competitive landscape changes. In this free-form (I call it "scramble") competition, your adversaries come out of the blue. Armed with a totally new product or process, they can change the playing field, and even the game itself, almost overnight.

CHAPTER 4

Scramble Competition

Beating New Competitors in a Wide-Open World

Technologically versatile competitors searching for new fields to conquer have radically changed the competitive landscape. As a result of their searching, which we described in Chapter 3, all of us in business now operate in a much different competitive environment than we did just ten years ago. I call this new environment "scramble competition" because it is disorderly, fast, and lethal. Traditional assumptions about business competition no longer hold true.

This chapter closely examines what scramble competition holds for any company, particularly those seeking to use technology to gain an advantage in today's marketplace. We see how competitors use this new environment to spring up and seriously challenge long-established players. We also see how your company can both protect itself and prevail in this environment.

Here are some examples of scramble competition, drawn from companies of all sizes:

- Paper manufacturers have found that Hewlett-Packard, whom many thought of as a manufacturer of instruments and printers, is a deadly serious competitor in the retail ink-jet *paper* business. Moreover, some of what H-P sells as "paper" is actually thick plastic film.

- Xerox finds that its most aggressive competitor is . . . Hewlett-Packard, again. HP is introducing a new line of multiple-copy printers aimed straight at Xerox's copiers.

- In a virtual end run around the big-city newspapers, Bernard Hodes, a Boston employment advertising agency, established an Internet-based service called Career Mosaic. It puts its own classifieds on the World Wide Web and so competes directly with the metropolitan papers' help-wanted advertising.

- Telecommunications giant AT&T issued the Universal card, a consumer credit card with no annual fee, to challenge financial institutions on their own profitable turf.

- Oil well service businesses are replacing two-mile lengths of coiled steel tubing made by huge steel mills in industrial cities with more durable, more flexible, stronger composite tubing from Fiberspar, the maker of windsurfing masts that we met in Chapter 3.

- Established makers of bulky, heavy cathode ray tubes used in televisions, computers, and monitors are under attack by more than 150 companies developing fourteen distinct technologies for flat-panel display screens.

- Amazon.com, the world's largest on-line bookstore, has rattled the major bookstore chains with its on-line book reviewing and ordering system. It was only recently that the bookstore chains themselves used superior inventory and ordering technology to overwhelm independent bookstores.

- Cable TV operators have been siphoning viewers from broadcast television for fifteen years. Suddenly they find Hughes's DirecTV and its minidish receivers taking away customers by offering great picture quality, lower cost, better service, and broader channel selection. Hughes is owned by General Motors.

In the immortal words of Butch Cassidy and the Sundance Kid, "Who are these guys?"

Surprise!

These upstarts are entering markets that are new to them. They are offering products and services that are equally new, all made possible by technological advances. On the hapless receiving end are strong, entrenched competitors surprised by how easily other companies use technology to restructure business competition. Scramble competition always seems to break the rules.

Scramble competition usually exhibits certain features:

1. Unexpected market participants begin competing against established businesses.

2. Small, emerging companies become leading market participants overnight, often with no warning to existing players.

3. Companies large and small use technology as a tool for entering new markets.

4. Technology is often combined in new ways with other resources, such as a brand name, distribution strength, or creative financing, to produce a fundamentally different product offering.

5. The new offerings compete very well against the old ones even though they don't match up in side-by-side comparison.

6. Traditional competitors seem unable to respond to the new competitive threats.

7. Buyers who try the new offerings seldom return to the old offerings.

Scramble competition grows directly out of technology-based strategy and is replacing "contest competition." Before we examine scramble competition in more detail, let's look at how we have been competing in that more structured system.

Contest Competition Is Comfortable

Contest competition in its purest form can be found in field sports, board games, and political elections. Contest competition exists only under rigidly controlled conditions. The competition is carefully structured, and by common consent the players compete within that structure.

Contest competition is comfortable because so much can be known in advance. It features a winner and a loser, so if you enter a contest competition against one other rival, one of you will certainly win, absent an occasional tie. That's comforting. Additionally, the field of play is clearly marked. It's comforting to know that there are boundaries and that when you cross them, the contest will stop. Rules like these give all contestants the soothing knowledge that competition is limited.

Contest competitors share ten common expectations:

1. The players are clearly identified.
2. All participants know whom they are competing against.
3. Everyone knows the rules and agrees to follow them.
4. The means of contesting victory are agreed upon in advance.
5. Participants who fail to follow the rules are penalized.
6. The contest has time limits.
7. The playing field has clearly marked boundaries.
8. The rules do not change during the game.
9. Apart from a tie, there is always at least one winner and one loser.

10 The contestants are playing to win the competition, not to destroy one another.

These assumptions bind the competitors together. Mutual acceptance of these rules makes the game or sport or business "fair" in the eyes of participants and observers. Contest competition is a friendly face-off requiring that contestants fundamentally agree on how they oppose each other.

Contest Competitors in Business

In business, contest competition usually takes the form of two or three major competitors battling for a share of a mature market. These competitors usually:

- Have been in the business for a long time.
- Know one another well.
- Tend to be about the same size, or are stable in size relative to each other.
- Keep score by tracking market share.
- Use essentially the same production technologies.
- Innovate at about the same stately pace, with no intention of reinventing themselves or their products.
- Compete by using the same well-known marketing and distribution strategies.
- Have reached accommodation with factors and entities that affect the business, such as unions, regulatory bodies, industry organizations, and government.
- Avoid price competition.
- Tolerate the presence of small contestants, whom they ignore as long as the little guys "stay in their place."

Contest competition is attractive to many businesspeople. The environment is harmonious. Managers can define and execute structured tasks, and they seldom face unexpected challenges. Knowing the competitors, the playing field, the calendar, the rules, the score, and the tools provides a sense of security. Managers know where they are and can usually find areas of good performance and measurable improvement to present in the annual report.

The orderliness of contest competition is so attractive that in many dealings between industry and government, the former uses the latter to preserve conditions of contest competition. For example, industry lobbies government to limit new entrants, regulate imports, maintain price supports, require minimum return on investment, or bar certain mergers and acquisitions. Unions use government to strictly limit the entry of new unions that might compete with them.

Contest competition in business used to be far more common. Before the major entry of Japanese competitors in the 1970s, the U.S. automobile industry was an excellent example. General Motors, Ford, and Chrysler, the Big Three automakers, competed for share using well-known tools such as national advertising, model year changeovers, cosmetic styling changes, and aggressive salesmanship by local dealers to sell essentially the same products. They were rarely, if ever, surprised by one another. None of them offered a car that was fundamentally different or a better value for its price. In another instance of contest competition, before the elimination of fair trade regulation in 1975 retailing was highly structured and predictable. So was the steel industry before the growth of minimills.

Despite its decline, contest competition still frames the thinking of the majority of managers. But those comfortable days are drawing to an uncomfortable close with the rise of scramble competition.

A Closer Look at Scramble Competition

Scramble competition is far less structured than contest competition. In fact, scramble competition's lack of structure is its defining characteristic. There are no boundaries to protect you, no rules to ensure your ultimate success, no time-outs. The game never ends. No one has agreed not to destroy you. Scramble competitors do not respect you. They don't want to play golf with you; in fact, they are intolerant of competitors in general. This is the type of competition that prevails in nature. Nature guarantees nothing. There are many possible outcomes and unlimited combinations of winners and losers. Anyone can play, at any time, using any tool that can be devised.

To see how completely natural scramble competition is, let's transport ourselves to the Serengeti. It's a pleasant morning, and a cheetah wakes up hungry. A herd of gazelles grazes nearby. The cheetah is well suited for this competition: extreme speed, quick reflexes, lethal claws, and sharp teeth. So he cuts out one older gazelle from the rest, runs it down during a zigzag chase, and kills it. But before he can claim his breakfast prize, he is chased away by hyenas that have been awaiting just this opportunity. The hyenas scavenge in packs, which the cheetah is unprepared to compete against, so he loses the second round of competition.

Then a lioness, who heard the noise of the hyenas and came to investigate, scares them off and eats her fill. She's so big, so strong, and so dangerous that no hyena—not even a pack of hyenas—can challenge her, and she wins this third round of the competition. After expending no effort in making the kill, the lioness simply shows up and claims the prize originally won by the cheetah.

After the lion leaves, vultures descend on the carcass. They, too, win and are later followed by other scavengers. The remains later provide meals for insects and even lower life forms. They are all winners. But the nominal "winner," in this case the cheetah who expended the effort to kill the gazelle, gets nothing.

This sounds a lot like business today, doesn't it?

In nature, the competition for nourishment is wide open. Any conceivable strategy for getting food can work. In business, a similar situation has shaped up. It is much harder to identify the competition. The boundaries and nature of play are uncertain, and the new competitors may have no stake in the business and don't care if they destroy a competitor. In scramble competition, strategy is wide open; creative tactics can often outdo superior resources, established position, experience, and execution of traditional moves.

Technology Creates Scramble Competition

Why are scramble competition and technology strategy closely linked? Please recall our statement in Chapter 3 that technology can go in all directions. When you take technology in new directions, you provoke new competitive situations everywhere you go. You become like Genghis Khan, living on horseback, sleeping in tents, and marauding the unprepared cities to the west. Technology creates unequal competition.

With technology, you upset every established competitor you encounter. Managers who develop or sell or use technology in new ways provoke scramble competition. Technology's marvelous variety allows managers to develop new solutions valued by new customers. When customers choose to work with you, they dislodge existing competitors who had these customers to themselves. Technology is a strongly destabilizing element in competition. Your choice is to use technology to your advantage or have it used against you.

What Is a Scramble Competitor?

A scramble competitor consciously and methodically:

- Uses technological advances as its primary competitive weapon
- Applies technology to deliberately create new markets and gain access to widely divergent markets
- Targets leadership and dominance in each of its chosen markets
- Aims to completely replace established competitors
- Takes a disproportionate share of the profits of its target businesses
- Uses its own advances and the work of others with equal facility

- Invests in technological advances with strictly business goals in mind

- Uses disciplined financial and strategic analysis to determine whether and how to enter a new market, and when to exit an existing market

- Is never satisfied with its existing position, size, profitability, and capabilities

Such businesses are dangerous competitors indeed. The largest scramble competitors are easy to recognize. Here are seven consistent scramble competitors to watch and study.

1. *Hewlett-Packard* is the ablest and most visible scramble competitor in the world today. Starting as an unknown instrument maker, H-P has used technology and the art of scramble competition to become a nimble, voracious behemoth, redefining competition wherever it chooses to go.

2. *General Electric's* competitive and managerial skills are legendary. GE uses technology as a competitive and business development tool in ways not widely appreciated—and does it well.

3. Perhaps no business uses as many different technologies as *3M* does to create market opportunities for itself. It is a uniquely determined technological competitor. More than any other, 3M takes many technologies in many directions. It has patent police worldwide, and one tests them at one's peril.

4. *Merck* is always thought of as simply a leading drug company, but did you know that it holds a hammerlock on supply of the liquid crystals that are used everywhere for watches, small displays, flat-panel computer screens, and calculators?

5. *Intel* may eventually try to reduce every function on earth to an element of semiconducting material on one of its microprocessors.

6 *Microsoft* writes code for anything Intel cannot miniaturize and includes it in one of its programs, which already run on 90 percent of the world's personal computers.

7 *Canon* turns its creative abilities to developing new imaging equipment for whatever new gizmo strikes our fancy.

If any of these Goliaths looks covetously in the direction of your markets, be aware—you are about to get scrambled!

Not all scramble competitors are large, multibillion-dollar corporations. A little guy with focus, momentum, and a new technology can damage your business as badly as a big guy, and the small ones are much stealthier. If size can't help you identify scramble competitors, how do you recognize them? One characteristic favoring scramble competitors is that you may *not* recognize them. In all likelihood, you may not have even heard of them before. You never thought they'd get into your business. After all, they have no history or experience in your business, so how could they possibly compete against you?

Knowledge of scramble competition helps you calibrate your competitive radar and defend yourself against these dangerous new competitors. No market is immune to this form of competition today. In fact, the absence of a scramble competitor represents an open invitation for one to join the fray.

Are You About to Get Scrambled?

How do you tell if you are under attack? Maybe your company is in the crosshairs of a scramble competitor right now. If your industry is characterized best by contest competition, then it's a candidate for a scramble competitor. Because scramble competitors can be so devastating—quickly grabbing high-margin niches by winning customers with new, high-value products—monitoring the horizon to find them must be a regular, active endeavor. This is best carried on by top management and by those close to the marketplace. The best initial defense is full alertness and anticipation. Andy Grove, chairman of Intel, was right: Only the paranoid survive.

Anticipating scramble competition requires you to dramatically change your definition of your competitors. You can no longer afford to focus only on the same industry players producing the same products for the same kinds of customers that you do. You must expand your definition of competitors to include any company that is somehow capable of meeting your customers' needs as well as or better than you can.

If you identify a new competitor, one that does business in numerous other markets not necessarily your own, you have located a prospective scramble competitor. The best scramble competitors, such as Hewlett-Packard and 3M, tend to play in many markets. If scramble competitors have begun to target your markets, they certainly believe they have something your customers will prefer.

Scramble competitors are willing to employ different strategic moves than you normally do. They prefer to compete in ways that put you at a disadvantage. If you sell on low price, they sell on high performance. If you sell a stand-alone product, they sell a system. If you sell service, they sell reliability that requires no service. If you sell through distributors, they sell direct. If you have a store, they sell on the Web. Such distinctive approaches to selling appeal to the most attractive customers on the basis of novelty as well as logic. Look for versatility in how scramble competitors conduct business.

Scramble competitors make moves that neutralize the investment of their new target competitors. If you sell powerful mixing equipment, they offer an additive that eliminates customers' need to mix at all. If your product is a finely ground powder sold by the bag, theirs might be sold as a liquid in returnable tanks. If you have invested in training distributors' salespeople, they put the information on a compact disc and give it away. They want to turn your investment into an anchor.

Often a scramble competitor's product radically changes the cost or performance picture for customers. If you perform a service, they might design an equipment module that fits right into the product's manufacturing line and performs your service function automatically on the assembly line. In all likelihood, a scramble competitor addresses customer needs by using an incompatible technology that you and your friendly contest competitors cannot easily duplicate. If some unfamiliar competitor approaches your customers with a new solution that entirely eliminates what you sell, you are about to get scrambled.

In one way or another, good scramble competitors satisfy some of your customers' needs better than you do. They are very systematic about this. Scramble competitors pick your market to pieces in bite-size chunks. They want to avoid a head-to-head battle, so they nibble away niches. They are judicious because they take your competitive clout seriously.

Keep a running list of proven scramble competitors from all the markets where you've seen them operating. Check this list often. Ask yourself, "Could any of these guys serve *any* of my customers with *any* product or service?" If you answer yes, watch them like a hawk.

How to Spot a Scramble Competitor

As a general rule, scramble competitors have little in common with you, so they're difficult to spot before they surface in your market. They attack from unexpected directions. However, you can identify them if they operate where you do. Scramble competitors can be spotted early when they:

- Make products similar to yours, but for completely different applications or markets.
- Serve your customers already, but with different products.
- Solve the same kinds of problems that you do, but for other customers and using different technologies (H-P watchers, take note).
- Produce very different products for other customers but share distribution structures with you.
- Produce radically differing materials that perform better than yours do in certain applications.

Despite these early warning signs, scramble competitors generally find you before you find them. Yet you must be able to find them as early as possible so you can assess them and defend yourself quickly. How can you find them? Here are some specific techniques:

- *Watch for nonindustry competitors.* Closely investigate any new nonindustry company that is luring customers away from you or any of your regular competitors. Keep your ears open for discussions of other companies among your customers. Monitor any outfit already doing business with your customers in any step of the value chain that includes your product. Keep an especially sharp eye on your most innovative customers and suppliers. Innovators are often the first approached by scramble competitors looking to test a concept or test-market a new application. Spot this happening and you've earned a six- to twelve-month advance notice.

- *Watch upstream and downstream businesses.* Your suppliers and customers could themselves become your next scramble competitors. Carefully watch the most innovative among your suppliers and customers for any sign that they are offering something that competes with you. Watch all the usual sources. If the most innovative ones are public companies, read their annual reports.

- *Watch trade show and acquisition activity.* At trade shows and industry conferences, look for unfamiliar names and faces. Take special note of atypical companies that get coverage in your industry's trade press. Watch industry participants doing something out of character. Monitor any merger-and-acquisition activity that affects your industry. A large company can buy a small one as an entree into your market, or use its technology in new ways or new places to do you in.

- *Survey customers biannually.* The biannual customer survey is an excellent tool for these sensing activities. Good customers talk freely about which new supplier is making overtures to them. Customers usually share industry chatter about new entrants or new ways of doing things. Be sure to include ex-customers in the survey to find out why they moved to other suppliers.

- *Monitor patent and trademark filings.* Note any company filing a patent or trademark naming your market or a niche of yours as an area of application. Even more alarming would be any company that cites your patents or "designs around" your patents. Watch for any company licensing a technology that competes against yours.

- *Teach your sales force what to look for.* Your customers and suppliers are the best source of information for advance warning of scramble competitors. Your sales force can be a good link to your customers for this kind of intelligence, but only if they realize the extent to which scramble competitors pose a threat. Since most salespeople believe in their products and companies, they often dismiss competitors who appear insignificant, tangential, or remote. Your sales force may look upon new, fledgling competitors with a jaundiced eye. Don't be misled by sales force disdain for the "oddball" or "toylike" products of scramble competitors. Those reflexive criticisms effectively camouflage scramble competition. Such dismissiveness allows scramble competitors to work for months or years before anyone takes them seriously. Lightweight products can become heavyweight threats.

Responding to Scramble Competitors

Finding prospective competitors is hard work. You're like a camper staying up all night watching for bears. Sometimes they never come, yet anticipating their arrival is easier than dealing with them once they arrive. After a scramble competitor arrives, you have an urgent challenge on your hands.

There are three rules to follow once you realize you face a legitimate scramble competitor. You should:

1. Seize the initiative.
2. Assess threats with a cool head.
3. Respond, don't react.

Let's examine each in turn.

Seize the Initiative

Scramble competitors have analyzed their target markets extensively. They know what they offer, why it's better, and which markets are most receptive. They think your customers are just fine—for them. Your being in the way is not of concern to them. They have sized you up as a target, and they have figured out a way around you. The best thing you can do is to take action instantly. Begin by assessing your situation and developing a plan of your own. Your new scramble competitor suspects that you'll do nothing about them. They have bet that they can move faster and more effectively than you can.

Once you have found that one or more scramble competitors are indeed after your customers, you must determine how immediate and how serious the threat is and then respond to it effectively. Delay in response never helps you with a scramble competitor, so seize the initiative. Don't delay acting because you think you don't know enough yet. The clearer the picture is to you, the more advanced the problem is and the less you can do about it.

Assess Threats With a Cool Head

The ideal frame of mind for assessing scramble competition is healthy paranoia, well short of panic. Once a scramble competitor shows up, coolly size up the situation. Analyze the competitor's size, position, resources, the nature of the application or market initiative that has been launched, and the time line along which the threat will materialize. Determine exactly what threat you face.

In general, *technological threats*—threats that might replace your technology in the market—are potentially the most serious. At a minimum, they erode your leadership and margins. At worst, they eliminate the basis for your business.

Technological threats come in two basic forms: encroachment and substitution. By definition, *encroachment threats* usually come from competitors in your industry; they seldom originate with scramble competitors. You know your direct competitors and most of what they're up to. Unless they get very creative and start scrambling, they probably pose visible and direct threats that you can address with equally direct means. They might use technology to offer improved quality or price or service, to which you respond in kind. Or you reply obliquely. You might counter lower price with better quality or service. If they have invaded one of your prize segments, you might retaliate by invading one of theirs. The nice thing about encroachment by contest competitors that is they have assets at stake in your industry that you can take hostage.

Substitution threats are much more serious. They typically originate with scramble competitors. They come from an outfit outside your industry that gets uncomfortably close to your technology, your applications, or your market. A substitution threat can eliminate you from a market or niche entirely. If you make polarizers for liquid crystal displays and Merck develops a liquid crystal that does not require customers to buy your polarizers, that's a fatal substitution threat. You can't really retaliate against Merck's substitution threat as you would against an encroachment threat. You can't fight Merck in its drug markets. You can't outspend them.

Still, you must not delay. If a substitute product adds value greater than yours and does so at significantly lower cost or higher performance, your customers are going to learn about it. Your scramble competitor insists on telling them. When they learn of it, that part of your business starts slipping away. You need to face the facts honestly and make your plans.

This kind of substitution threat highlights a frustrating aspect of scramble competition: Scramble competitors come out of left field, and you often can't do anything about them directly.

How quickly the business slips away depends on many factors, including the usable life of the product, the buying cycle, and how entrenched your product is in customers' operating systems. Given these criteria, if you, for instance, manufacture building material for public-sector construction projects, substitution threats probably take longer to materialize than if you make office products. If you make thermal fax paper, you face an immediate threat from plain-paper fax technology.

The life of the product and the buying cycle determine the tempo of the market. *Tempo* here refers to the speed of adoption by the various buyer segments along the adoption-diffusion curve. In assessing the speed at which a new, truly competitive technology could substitute for yours, you must consider your customers in terms of this curve. If your customers are innovators and early adopters, you face a more pressing near-term threat than if your customers are late majority buyers. The adoption profile of your customers determines how much time you have to launch a response.

Next, you must examine the implications for your company's market segments, particularly the high-price, high-margin niches you prize. These are what often attract scramble competitors. It is also where they can do the most damage, particularly if you have not established a strong leadership position. A niche competitor with a high-value application essentially puts you on the receiving end of a TechnoLeverage play: You can lose valuable business on the right-hand end of the lever. Victims of scramble competition risk erosion at the high-value, high-price, high-margin end of their business. They wind up with the rind of the watermelon—the lower-value, lower-price, lower-margin markets—while the scrambler takes the sweet, juicy center. This is, of course, a serious threat that must be met aggressively and decisively.

Respond—Don't React

Once you determine that you face a true threat from a scramble competitor, you must respond to it. A response differs from a reaction in that a response is reasoned and strategic; it has greater probability of addressing the actual threat. A reaction, on the other hand, is reflexive, unthinking, unplanned, and likely to be ineffective in addressing the threat. If we're going to play a new game, we must keep our wits about us.

The starting point for your response must be your analysis of the threat. The better the information you have on the threat and the better your analysis of this information, the better you can direct and calibrate your response. But this is only half of the information you need. In such a situation, detailed knowledge of your customers—how they use your product or service, how they view it, their plans for the future, their purchase-approval procedures, their position on the adoption-diffusion curve—helps tremendously.

Three Strategic Responses

There are three fundamental strategic responses to scramble competitors. In declining order of desirability, they are:

1. Change the way you do business.
2. Defend your strongest niches.
3. Exit the business if you cannot succeed.

Let's discuss each of them.

Change the Way You Do Business

Changing the way you do business is the best strategy to adopt if you are willing and able to invest money, time, effort, and other resources to match the scramble competitor's value. Your willingness should be based upon calculated appraisals of the threat and of your chances of successfully meeting it. This is the boldest, highest-commitment response to a scramble competitor. It calls for an all-out effort, not a piecemeal approach.

In the context of this strategy, there are several moves you can make. You can radically change your product, enhancing it to match or exceed the value of the new offering. Or you can sharply lower your price to reflect the new, relatively lower value of your offering. This often works in the short term because it slows the migration of your customers to the competitor's product. You can also adjust your terms of sale by offering improved service or better warranties or alternative financing such as a leasing plan.

This strategy eventually fails if the scramble competitor keeps improving its offering. But improving your product or reducing your price can buy you time. Your most loyal customers and your sales channels will stick with you for a while. Others may give you the benefit of the doubt, waiting to see what you come up with as a longer-term response.

You can also try to shape industry standards around your product. Although it may be late to try this once you face a scramble competitor, you can still benefit to the extent that you further bind some of your customers to you. You can introduce an enhancement of your own that depends upon your technology, or you can achieve some long sought-for standard the market has awaited, such as compatibility with overseas standards.

You can make a deeper commitment by purchasing or licensing the technologies you need so as to compete with the new entrant directly. The success of this strategy hinges on the degree to which deal making helps you exceed the improved value of the new offering, the speed with which you can bring your version to market, and your success in selling your response product to existing and new customers.

At a still deeper level of commitment (and profound admission of the competitor's superior value), you can try to acquire the upstart competitor. This strategy is tricky since it calls for a potentially large investment and for management commitment to accommodate a different style. Numerous financial, cultural, legal, and operational issues come with acquiring a competitor, particularly when an upstart is acquired by a traditional company. You could acquire a competitor of the threatening upstart, but that adds the risk of buying a me-too firm.

When you decide to respond to an upstart competitor by changing the way you do business, you are refusing to give in and choosing to be flexible; both are healthy.

Defend Your Strongest Niches

This is the best strategy to adopt if by objective measures your product remains the best solution for some markets but a weak offering for others. You must carefully assess which applications are defunct and which serve markets where you still have the best technology. This response implies yielding some territory to the scramble competitor.

The success of this strategy hinges upon two conditions. First, it requires honest, accurate appraisal of which markets you can serve and which ones you are destined to lose.

Second, you must live without the territory you are yielding. If for you it is a small, slow-growth, low-margin segment of your customer base, you're not yielding much of importance. Unfortunately, although it may be small, it may not be low-margin, for few scramble competitors pursue a small, low-margin market—at least not ones that they believe will remain low-margin for them. Scramble competitors tend to target high-margin business.

Typically, a scramble competitor begins by entering a carefully chosen niche of what for you is a large, slow-growth, often low-profit business. The scrambler does this by applying its technology to a costly, persistent problem in this niche. In doing so, the competitor uses its product to move that niche from the high-volume, low-margin, commodity end of the technology applications spectrum back up to the exotic or specialty zones. This first niche is often the thin end of the wedge that an upstart uses to enter a larger market.

When Honda first leveraged its engine technology into the U.S. market in the mid-1960s, it targeted the motorcycle market. Within motorcycles it targeted the small, lightweight end, where none of the established competitors had any serious interest. I remember clearly the announcement that Honda's then new Marysville, Ohio, plant could also produce automobile engines. People were puzzled by this announcement. Automobiles by Honda? Automobiles needed big engines. Later Honda did enter the automotive market, the same way it entered the motorcycle market, by stalking the low end where the Big Three took few if any profits. Detroit thought, What was there to fight about? This brings up TechnoLeverage Tip Number Four:

> ▶ ***TIP NUMBER FOUR:* Lack of interest by existing suppliers is the best possible invitation to a scramble competitor.**

If this describes your industry's attitude, watch out.

Because this is the thin front edge of a potentially very thick wedge, yielding this first piece of territory can cause real trouble down the road, particularly if more than one scramble competitor is probing the perimeter. A traditional company beset by several savvy scramblers can quickly see its markets, sales, and earnings shrivel if it yields territory to them one by one.

But trying to defend every piece of your territory against existing and scramble competitors at the same time proves fruitless. As the military has observed, "He who defends everything defends nothing."

If the market you serve well moves slowly and your application has a long and useful life, you could have a solid, profitable business for many years to come, as well as plenty of time to work out your next move. After all, the U.S. Postal Service survived the success of Federal Express and even drew a few ideas from it.

Exit the Business If You Cannot Succeed

Exiting is the strategy to employ if a thorough analysis of the competitive threat reveals that you cannot compete effectively. If you conclude after careful analysis that you cannot change the way you do business, that you are unable to face down or co-opt the challenger, and that you cannot hold on to enough markets to make continuing worthwhile, why not sell while there's still something to sell?

In other words, if neither of the first two strategic responses is viable, exit the business. Face it: Businesses can die. Some get murdered by new competitors with unbeatable offers. Some bleed to death from the loss of their best niches. If you can't lick 'em and you can't join 'em, what else can you do but sell?

The alternative is to take yourself, your employees, and your shareholders though a painful, protracted period of declining sales, eroding markets, and evaporating capital. Many management teams choose this option simply because it is so hard emotionally to throw in the towel. A strong contest competition ethic of not giving up, together with a fear of being labeled a "quitter" by others, also supports management's decision to prolong a losing battle. But at the root of such folly lies a failure to face reality squarely and deal with it resolutely. Ideally, the business's managers, employees, and shareholders should all go find something better to do with their time, money, and talent. They should do something that has greater value to the market.

It takes brass to pull off a graceful exit and sell out at the right time. If a scramble competitor's handwriting on the wall is accurate, it may amount to an early warning of where the whole business is headed. Once a business starts an inevitable decline, the sooner you exit, the better. It is the buyer's responsibility to exercise due diligence and to buy at a price that reflects the true value of the business. Could Otto Eckstein, the Harvard professor who founded Data Resources, have seen the demise of timesharing in the rise of mini- and microcomputer technology when he sold the firm to McGraw-Hill in 1979? Did Hanson PLC foresee the inevitable fall of the typewriter when it spun off Smith Corona in the early 1990s? Did Leonard Florence foresee the complete saturation of the market for molded plastic chairs when he sold Syroco in 1996? Who knows, but their timing could not have been better if they had. Smart businesspeople are completely honest with themselves about their prospects, and when the time comes to sell, they sell.

A graceful and profitable exit can often be made gradually. You do not have to exit an entire line of business all at once. Instead you can time exits from various niches—spinning off divisions, raising prices gradually, closing down units—and reallocate resources accordingly. Echoing the message of Chapter 3, the more numerous the attractive alternatives you have toward which you can reallocate the freed-up resources, the better off you are. Developing these alternatives, an essential part of technology strategy, enables you to perpetually renew the business—even in the face of scramble competition.

Start Scrambling

As in most competitive situations, the best defense is a good offense. To succeed in our environment of scramble competition, become a scramble competitor. Because they work at the leading edge of their field, scramble competitors often become leaders in their technology, applications, and markets. Metromedia Technologies is now the unquestioned industry leader in huge-format digital printing for billboards and large displays. As we see in the next chapter, such leadership positions provide excellent vantage points from which to deal with all competitors. Among the major benefits of leadership is the fact that the leader, by definition, gets to set the standard, particularly in a technology or application.

Finally, scramble competitors innovate constantly, even if it means that their own products become obsolete in the process. Andy Grove calls this "ripping up the road behind you." It makes life very hard for scramble competitors as well as for anyone trying to play follow-the-leader.

The message is simple: Anyone can start to scramble. In this environment, everyone had better.

Review and Preview

In this chapter, we've seen that technology destabilizes competition. One result has been what I call scramble competition. You can either become a scramble competitor or risk becoming a victim of one.

To become a scramble competitor, use TechnoLeverage Tip Number Four—*Lack of interest by existing suppliers is the best possi-*

ble invitation to a scramble competitor as a divining rod: Find a high-potential (for you) niche in a new market ignored by the existing players, and target it. Apply your technology (or someone else's technology) to go after the sleeping players' high-margin niche. Find a way to solve a customer problem or serve a customer need that is not currently solved or served, and do it with technology. Crank up the pace of innovation, add great value, price high, and aim to completely replace the established competitor in your targeted niche.

To recognize a scramble competitor, watch the market closely for the unusual player on your turf. Pay close attention to the threat of substitution. When you see customers trying new, unorthodox solutions, you may be about to get scrambled. If that happens, basically you have three choices: Change the way you do business, defend your strongest niches, or, if those two strategies fail, exit the business.

In the next chapter we look at two logical extensions of successful scrambling and achieving TechnoLeverage: technological leadership and market dominance. A company does not necessarily have to become the leader as a result of TechnoLeverage, but the rewards can be commensurate with the effort leadership demands.

CHAPTER 5

Ahead of the Pack

Taking and Holding the Lead

To get the full advantage of TechnoLeverage, you must be a leader. The question is, leader of what? In this chapter, we identify the areas where it is important to lead, and we examine how to make your organization the leader and keep it there.

In the corporate strategy world of TechnoLeverage, there are three things a company can lead: a technology, an application, and a market. Let's define what leading each of these means, using as a real-life example blue-light-emitting laser diodes. Lasers happily come in reds and greens, but a good blue laser is very hard to achieve. That's too bad, because blue lasers would have many useful applications.

Technology Leadership

In the technology realm, the definition of *to lead* is to have a margin of advantage, to have superiority, or to be foremost in some respect. That's what people pay for when they buy technology: They want advanced performance, something that's better. A technological leader may have broad superiority within an entire technology, just as Xerox's xerographic technology leads carbon paper and mimeograph.

Technological leadership may also be narrower, lying along some dimension of performance. The technology of the blue-light laser diodes is led by a small Japanese company, Nichia Chemical Industries. Nichia's diodes are very blue and exceptionally small, which allows high performance and fine-grained resolution wherever they are used. Nichia's leadership of this technology refers to its margin of advantage, the superiority, the "foremostness" over other approaches, that customers seek and that companies deliver with technology. Xerox is trying to come up with a better blue-light laser diode than Nichia has. Xerox is large, powerful, and well known, but in this technology it is not a leader. It is barely even a follower.

Application Leadership

In the realm of applications—commercial use of a technology in a product—the relevant definition of *to lead* is to direct or to guide. When you are the leader in an application, you guide the rest; you set the standard. You fiddle the tune that others must dance to. Application leaders are the pathfinders, the Daniel Boones who go out into the new land and say, "We'll call this place Kentucky," and then everyone calls it Kentucky because the one who got there first called it Kentucky. Leadership in an application means solving a problem so well that you set a standard that others must follow.

Nichia's diode is exceptionally small and effective. It could soon be used in two separate applications: compact disc production, and high-resolution copiers. Nichia's laser diodes are suitable to either application, but Nichia has not yet commercialized its technology in either one. Xerox can't participate in any applications yet because its attempts at making blue-light laser diodes aren't satisfactory. Neither company is a leader in these applications for blue-light lasers because neither is a commercial contender.

Market Leadership

In discussing leadership of a market, the reference point is other suppliers, other participants. The definition of *to lead* that is relevant to markets is the one you may have thought of first: to have first place or to dominate. Regardless of its limited progress with blue-light lasers, Xerox is the market leader in larger-size office copiers. In copiers, Xerox is everywhere; it has huge revenues, many customers, many technologies to draw upon, and many products. Nichia is not a market leader in copiers or CD production equipment.

TechnoLeverage leadership differs from the market leadership battles waged by Coca-Cola against Pepsi, or McDonald's against Burger King, or Hertz against Avis. The pursuit of leadership through technology strategy deemphasizes the traditional battle for market share per se, which is marked by outthinking and outspending competitors on brand advertising, product promotion, and distribution. In contrast, TechnoLeverage leaders use technology to redefine the game, redraw the battle lines, and change the playing field and the standards in their favor. Properly applied, technology gives you brand-new fields to lead. With TechnoLeverage, companies fight not in neat lines like redcoats, but like American colonists—from behind trees and stone walls.

Let's look at some of the chief characteristics of leaders from a technology perspective. As we do so, consider how your company and its people rate as leaders right now, and how you might encourage them to improve. Keep in mind that leadership in the context of TechnoLeverage applies to any company that wants to use technology for strategic advantage.

Characteristics of Leaders

TechnoLeverage leaders are *determined*, especially about leading their technology and their applications. They want to be leaders more than anything else. They can't stand to think of someone being able to do something better than they do. They'll endure any adversity to achieve and maintain a leadership position.

Being out in front, the technology leader must be *independent* and *self-reliant*. Technological leadership is not for conformists because acquiring a leadership position takes you to unexplored places. Applications leaders are often criticized, particularly for offering an unusual product. If you are going to push your business to leadership, you are going to encounter criticism, second-guessing, loneliness, and misunderstanding. The rewards of technological leadership outweigh these discomforts.

I emphasize that leaders are *different* because technology allows a company to differentiate itself and its products from others; this is one of technology's chief advantages. Technology is a remarkably effective differentiator. It enables a company to progressively differentiate itself by making its products increasingly distinctive. With this distinctiveness, TechnoLeverage competitors stake out territory in which to lead. Leaders must be comfortable with the scrutiny that leadership invites.

There is *certitude* in leaders. They may turn out to be wrong, but they are always sure. Technological leaders display certainty because they are *visionary*; they see clearly what technology makes possible, and they urgently want to make that vision a reality. This takes absolute certainty. Leaders' sureness reflects a belief in their vision, even when the vision is outside their peers' comprehension, as was FedEx's implied statement back in 1976 that "Yes, we can fly all the urgent packages to Memphis and then fly them out to their destinations by the next morning."

Leaders are *goal-oriented*. Whether they work in R&D, product development, or sales, they focus intensely on getting the job done. They are single-minded about accomplishment. This focus can give rise to a militarylike concentration on effectiveness. However, effective leaders—particularly when they are seizing the lead—can also be inefficient and messy. Still, they achieve their goals.

The sureness, effectiveness, and occasional inefficiency of a technology leader recalls America Online's mid-1990s drive to become a leading provider of access to the Internet. AOL advertised heavily, carpet-bombed the nation with complimentary disks, and offered unlimited Internet access at extremely competitive rates. Did AOL incur astronomical costs with this program? Definitely. Did it lose money in the early stages? Yes. Did the company anger users when it did not have enough capacity to serve all comers? Absolutely. But AOL wanted leadership and got it.

All leaders have doubts, but they tend not to share them. They make up their minds, look forward, and pursue their goal. They favor *speed*. They know TechnoLeverage Tip Number Five:

> ▶ ***TIP NUMBER FIVE:*** **What one engineer can create, another can copy, so keep innovating.**

Thus they work in an unremitting state of controlled panic lest someone else beat them. Their sense of *urgency* echoes the thought from Chapter 3 that impatience is a virtue in technology strategy.

TechnoLeaders are *flexible* as to means. They see technology as a tool, as a means to an end rather than as an end in itself. They try one approach, and if that fails, they try another, to the point of solution or exhaustion. Recall Thomas Edison and his pursuit of the right filament material for incandescent bulbs. Legend has it that he tried hundreds of approaches before finding that tungsten wire in an argon atmosphere did the trick.

Finally, leaders are *choosy*. Leaders fight on their own terms, not on the enemy's. The best technology generals fight where they want to fight, in a place where their troops have the advantage; they avoid fighting on the enemy's turf and terms. Sound technology strategies for gaining leadership are difficult to respond to. Companies that gain leadership with technology try to pick a battleground and tactics that can't be responded to "fairly." If you're Motorola, how can you fight Qualcomm's lead in spread spectrum cellular telephone handsets? Qualcomm's code division multiple access (CDMA) is a better-performing technology, a better solution. When Intel designs your adjacent functions right onto its chips, how can you fight that? It is extremely difficult to answer such moves. TechnoLeverage leaders are therefore choosy about their struggles.

As we look at strategies for seizing and holding the lead, note that many leaders' competitive moves are largely *creative*. They strive to add value in new ways. Technology leaders are also *competitive*; they revel in the game aspects of what they do. They want to be first. They want to beat you. They don't want you to be able to fight back.

The Advantages of Leadership

As discussed in Chapter 4, contest competition centers on structured competition for size and share. The expected outcome features one or two big winners; the rest of the competitors are small and have low share, so they are "losers." Contest markets stay static because it is hard for ordinary companies to enter them and try for leadership. That would be disruptive, so all contestants try to prevent it.

But with TechnoLeverage, most companies can achieve leadership somewhere if they choose to do so. They simply need to solve any of mankind's limitless problems by drawing from technology's limitless advances. They can lead if they make the connection between problem and solution, achieve the advance with technology (theirs or someone else's), develop the application, and enter the market. They don't need permission.

Is being a leader worth the trouble of finding those matchups? Absolutely! Being the leader confers five valuable advantages in today's environment. Leaders get to do the following:

1. Set the standards and the tempo.
2. Occupy a prime position.
3. Secure the reference customers.
4. Grab the early high margins.
5. Dominate through their experience.

Set the Standards and the Tempo

First, the technology leader gets to set the standards for those who follow. The leader gets to establish customer expectations for the product or service by first structuring the offering. EDS structured the competition for large-scale information systems integration: outsourcing it. FedEx set the standard for fast delivery: overnight. 3M set the standard for digitally printed full-color truck graphics: completely weather-durable with seven-year warranty. Such companies become strongly dominant because every follower is judged by the standards set and expectations created by the leader.

Often the leader literally sets the standards for their products. The first provider of a new technology has to establish benchmarks and testing standards for producing the product and communicating its specifications to buyers. In many cases, the leader is able to stipulate these tests and quality standards and determine the criteria. This is a potent advantage with technology products.

The leader also sets the tempo of business competition. The tempo affects the *speed of adoption* throughout the entire market—the speed at which various buyers, such as innovators, early adopters and so on, buy—and the *speed of new application development*. From the leadership position, you can call the tune the followers must dance to: If you polka, they must try to polka. Fidelity was for years the technical leader in the mutual fund industry, pouring hundreds of millions of dollars into communications and computer technology and forcing its competitors to spend faster and faster to keep pace.

Occupy a Prime Position

Second, the leader usually has the chance to occupy exclusively an attractive physical position in a market or an application of a technology. From this chosen point, the leader gets to play king of the mountain and compete with the latecomers. Catalina Marketing, of Petersburg, Florida, which makes point-of-sale printer systems for customized coupons handed out in grocery stores, got to the checkout counter first. Catalina got the critical position. There aren't going to be two such systems on a counter, so a follower who wants to enter later has the hard job of dislodging Catalina.

Secure the Reference Customers

The leader's third advantage is getting the "reference customers." These are the early customers whose experiences are a critical sales tool for the leader in selling its new application to others. The newer the application, the more important these customers are. The most critical reference customers are innovators. Once you have them as customers, you can say to all other customers, "Look, we served ZXY Corp., the acknowledged leader in your industry. It worked for them, and it will work for you, too." The experiences and reference value of innovators and early adopters, and their opinions of you as the leading supplier, count for a lot with subsequent customers. If you lead the way into a market and get the earliest customers, you can lock in these referrals. Then you are the most experienced supplier when the more conformist early majority customers begin their purchase evaluations.

Equally important, first customers often become the leader's customers for life, especially if the application satisfies them and sets the standard for years. Customers who adopt the application and build it into their operations often see no reason ever to desert their first supplier.

Grab the Early, High Margins

Most important financially are the early, high margins that only the leader can achieve. The technological leader can practice nearly pure value pricing. Without competitors, the leader charges whatever he can get. The leader got there first (with the technology, into the application, or to market) and, as long as he can maintain some degree of exclusivity, he can charge high, value-based prices.

Many business executives shy away from developing technological advances in new directions, thinking they are expensive. But early advances in new directions are usually the cheapest and most accessible ones to develop. The first ore out of a mine is usually the cheapest to extract. As the miners take out the better-grade ore, the remaining ore requires greater and greater expense to extract and refine. The same is true with technological advances.

In contrast, PairGain Technologies, based in Tustin, California, has a new technology for increasing the transmission capacity of subscriber loop telephone wires, the "last mile" into your home from the central office. PairGain's operating margins are 28.5 percent and likely to keep rising. There aren't effective alternatives yet, and the cost of PairGain's solution is cheap compared with the cost of rewiring the subscriber lines. Leaders like PairGain can and do exploit the supply-and-demand mechanism that is at work when an advance adds great value and there are no competitors. Leadership should be a high-margin play. The goal of leadership in technology strategy is to take full advantage of the higher-margin zones of the TAS. These high-margin zones provide the financial rewards of technological innovation.

Dominate Through Extensive Experience

Finally, the leader, the first one into an application, always has more experience than the followers. Cumulative experience is a tremendous advantage in high-technology applications and markets. In contrast to popular lore, the TechnoLeverage pioneer is the one *ahead* of the guy with an arrow in his back.

A business can repeat the process of finding newer, higher-margin applications and enjoy a second-order-experience effect. Intel repeatedly designs new generations of chips. Through repetition, it has acquired experience in reconfiguring succeeding generations of microprocessors, and this makes Intel a difficult adversary.

Any Company Can Lead

Although some of those mentioned above as leaders are in high-technology markets, any company can use technology strategy to pursue leadership. NicSand Inc., a Berea, Ohio, maker of sandpaper, decided to grab leadership in a segment of the abrasives market. Sandpaper has literally been around for centuries. The Cleveland *Plain Dealer* reports NicSand's founder Jim Sandusky as saying, "The industry was asleep for decades. It just needed someone to wake it up." Founded in 1982, NicSand focused on the automotive market and decided to seek leadership by establishing a technological advantage.

NicSand develops products aggressively, often pursuing high performance along a certain characteristic of abrasiveness: fineness. Its patented Power Polish is an extremely fine abrasive applied to a power tool attachment. The benefit is a very fine polishing effect, with the ability to remove very shallow scratches in metal, plastic, and other materials. Another NicSand product, Power Wool, an extremely fine and durable type of steel wool, also provides new levels of polishing performance. The company regularly introduces new items.

NicSand's development efforts are augmented by the company's willingness to use others' technology. One early application was using Velcro to fasten sandpaper to power tools, an innovation that kept the sandpaper in place without tearing it as competitors' glue, screw, and clip arrangements did.

Part of NicSand's strategy was to seek leadership in a sizable niche—the consumer sandpaper segment, which includes sandpaper for wood and automotive applications. The company's major competitors, 3M and Norton, huge companies with large abrasives divisions, focus on the industrial market, which accounts for 75 percent of the $1.2 billion market. As of 1997, NicSand had 61 percent of the retail automotive sanding market and a good chunk of the home wood-sanding market.

Sandusky's company is an excellent example of TechnoLeverage leadership. The company's motto is, "We do what others don't." This kind of company and its CEO would be leaders in any business they pursued. They applied technology to a sleepy, mundane business to achieve leadership in a selected area.

Does the Biggest Get to Be the Leader?

Many of us associate leadership with large size. General Motors, Exxon, and Merrill Lynch are huge organizations in terms of assets and sales and head count. Sometimes the big ones are leaders in profitability, too. But you need not be the biggest to be the leader. We're all aware that the biggest outfit in an industry is seldom the fastest. Speed of technical development and effectiveness of market development, rather than size, determine the leader, particularly with technology strategy. Size and length of time in a business matter little to innovators and early adopters who need a solution to a large and costly problem. If you have the solution, they buy from you to solve their problem. They'll worry later about service in remote locations and how long you'll be around.

Another reason that size is less important is that early-stage, high-value-added, technology-driven applications are frequently self-financing. A company can pursue a make-a-little, sell-a-little strategy among early buyers. Or the application may proliferate so fast at such high margins that the returns come quickly enough to finance growth. An outfit with such a profitable application builds a war chest that can be used to grab the lead. Admittedly, there frequently are high R&D expenditures to amortize and recoup. However, the cash flow from a successful new application or new market can carry a company all the way to a stock exchange listing.

Even though size is not a key factor in achieving leadership when you are selling a new solution or advanced performance, it can sometimes *mean* leadership. Size matters more in consumer markets because late majority and laggard buyers, who make up half these markets, need many forms of assurance before purchasing.

Large size is a huge advantage in the last stages of the product life cycle—in maturity, saturation, and decline—and in the commodities zone of the TAS. Markets are large in that zone, and economies of scale in financing, operations, purchasing, distribution, and management often determine leadership.

Large companies can pursue leadership in specific technologies, as we've shown with Metromedia, Citibank, and Intel. Any time a company, large or small, focuses intensely on adding value with technology, it is on the path to achieving leadership through technology strategy.

Proven Strategies for Taking the Lead

Assuming a company does not have a leadership position yet truly wants one, what are the best strategies to employ? There are five steps to take:

1. Target markets that are high growth for you.
2. Seek an opportunity to make a technological advance.
3. Plan intensively.
4. Get started and keep moving.
5. Fully exploit the opportunity.

Target Markets That Are High Growth for You

Of course you can find high growth in markets that are growing fast. That needs little explanation. But you can also find fast growth for your company in a big, slow-growth market, as NicSand did. What's the harm in targeting a few niches that aren't moving so fast, so long as you can speed them up yourself? There are many such markets, and even a $50 million to $200 million business can find attractive growth making such a niche into a fast-growing market for its products.

Finding markets that are high growth for you has several implications. For openers, this mission keeps you from using your current market as a crutch. It prevents your saying, "We're not in a growth industry." That kind of talk gets you nowhere. Nobody would call the sandpaper industry high growth! But through creative problem solving and innovative application of technology, NicSand created high sales growth for itself in a slow-growth market. NicSand is growing at 37 percent annually, but you won't read cover stories about sandpaper as a hot, high-growth market. NicSand makes its own opportunities.

Discovering a market comes from paying attention to what others can't see, or to what is seemingly too small or too unattractive for them to pursue. Don't dismiss a market just because it appears small. There are many $10 million markets that offer 70 to 80 percent gross margins, particularly in specialties such as chemicals and composites. Of course, attending to smaller markets calls for judgment: There are markets that are inherently too small, for example, the market for fluids used as imaging contrast agents in ultrasound diagnostics. This is a small pond no matter what kind of frog you are.

The beauty of TechnoLeverage as a strategy is that you can individualize it to fit your company. You can design your strategy around your company, your people, your technology, your capabilities, and the problems your prospects and customers face. The individuality and independence of this approach stimulates leadership.

Bob Ennen founded NN Ball and Roller Company, in Erwin, Tennessee, to make high-precision bearings for specialized uses such as oil drilling. He applied specialized finishing technology, his own and others', to make uniformly round, smooth ball and roller bearings. Then he targeted as customers businesses making their own bearings, offering to outsource their production at lower cost and higher quality. He has built a business around his vision and his targeted industry.

Seek an Opportunity to Make a Technological Advance

If you see an opportunity to make some nontechnology move—cutting your costs, improving your merchandising or whatever—take it. But persistent high margins and fast growth result from applying technology to solve a problem or to create new markets, as microcomputers did. Technology is the differentiator that produces leadership.

The search for differentiation is why GM's and GE's railroad locomotive divisions have labored so mightily to introduce alternating-current (AC) technology to diesel locomotive traction motors. Traction motors have a difficult life. They work doggedly hour after hour, applying vast amounts of turning power, slung on an axle of a huge locomotive, pounding along without springs or shock absorbers, in snow, rain, dust, and sand. DC traction motors break down often and are expensive to repair. AC technology motors are much simpler, more rugged, and easier to repair. So AC traction motors would be a huge advantage to the railroads that adopt them, and to the locomotive company that supplies them. GM and GE have each pushed the AC advance hard and introduced it into service, each trying to differentiate itself from the other with a technological advance. We'll see in years to come whether two companies choosing the same technology to push ahead are making the right competitive move.

Be ready, willing, and able to apply anyone's technology to the problem. You sharply limit your opportunities by thinking you must invent it. Metromedia used someone else's ink-jet technology to print billboard images. NicSand used Velcro, sold by Velcro Industries BV, based in Amsterdam, the Netherlands, to fasten sandpaper to power tools. Intel integrates others' manufacturing technologies. These companies take whatever is available; acquire it through license, purchase, alliance, or other arrangements; and apply it to their customers' needs.

In seeking opportunities to make advances, remember the steps from Chapter 3. *Scanning* alerts you to opportunities. *Elaboration* helps you find specific ways to capitalize on these opportunities. *Balanced evaluation* helps you choose the most profitable way. Achieving leadership depends upon the breadth and intensity of the customers' need, the uniqueness and actual performance of the application, and the characteristics and competence of the company that brings it out.

Plan Intensively

Having identified the market and the application, you must then turn to a more traditional business activity: planning. You need a development plan; a competitive plan; a production plan; a financial plan; and sales, marketing, and distribution plans. Once you've selected a target application and market, planning makes the technological solution a business reality.

Great military leaders examine the facts and then plan carefully. Many of the great generals were outstanding topographical analysts. George Washington, Ulysses S. Grant, and Napoleon were all skilled geographers. Washington was a surveyor. They all studied their maps until they understood thoroughly the territory, their options, and their opponents' options. These leaders didn't "Just Do It."

It takes time, effort, and talent to assess the costs, risks, returns, competition, production and technical requirements, and likely problems and then develop contingency plans. Successful companies know they are treading on someone else's turf, markets where they know less at the outset than the other guys. They correct this imbalance with aggressive study of the terrain ahead. Taking the lead demands valuable data, serious analysis, and thorough planning.

Get Started and Keep Moving

Once the decisions have been made and the plans completed, it is time to act. As noted in Chapter 4, many companies that seize technological leadership are scramble competitors. They've done their homework on materials, competition, customers, and alternative solutions. They have the facts. They know they will become leaders only by taking action. In using technology, nothing can be gained by delay.

Decisive action sometimes calls for the ability to let go of the past. A company might want to hang on to its current business while developing a new area. It fails to commit enough resources to take the lead in the new area; it hedges its bets. It assigns people to the new effort part-time, asking them to capture a leadership position in addition to their "real" full-time job. This split focus is self-defeating. NicSand did not try to focus on both the commercial market and the consumer market. Rather, it pursued consumer market niches and left the industrial market to 3M and Norton.

Fully Exploit the Opportunity, and Overwhelm Potential Competitors

Whatever need you are addressing, address it as fully as you can. Leave no gaps that can be exploited by competitors. If the product must come in five sizes, make it in five sizes. The better your cover-

Figure 5.1: "Burma Shave" Wisdom for Technology Leaders

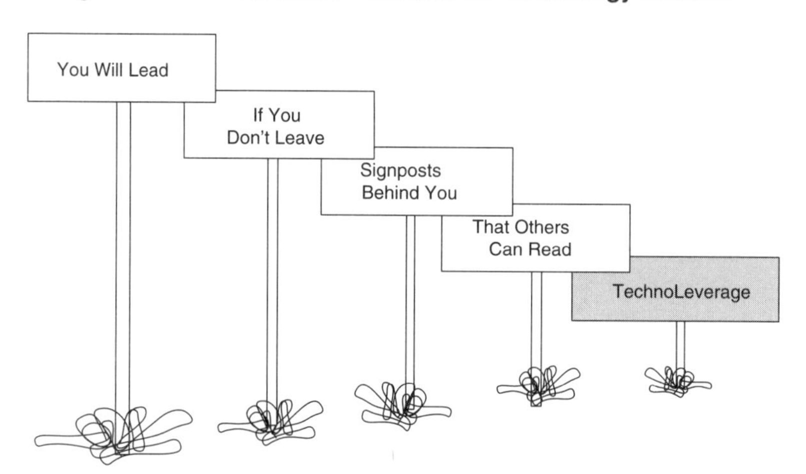

age of the opportunity, the better you will ward off competitors. Remember that the primary strategy of followers is to find and fill gaps in the offering of the leader. As the leader, fill as many gaps as you can. Based on the old (OK, very old) Burma Shave highway signs, Figure 5.1 advises you not to point out opportunities to competitors.

Leaving gaps is a danger inherent in getting your application out fast, so guard against it. Strike a balance between getting it out fast and leaving no room for a competitor. If no such potential competitor exists (and it's tough to be sure that this is so) and the need for the application is intense, get generation one out fast, leave a gap or two, and fill them in generation two. Recall that Hewlett-Packard and Hammermill did this with ink-jet paper.

If you fully satisfy the customers' needs, they won't think of going anywhere else. They won't need to. They become your reference customers and, hopefully, customers for life. If you scare off potential competitors, your customers won't have anywhere else to go anyway.

It takes commitment and resources to overwhelm an opportunity and intimidate potential competitors. Far too many companies starve new products or activities because they are afraid to commit to them. As a result, they ensure failure by underresourcing.

The economic forecasting company DRI/McGraw-Hill is a good example of failing to commit resources to new products and services. Originally called Data Resources, Inc., DRI was the first economic forecasting enterprise to become a substantial business; it did so through timesharing. Timesharing makes software programs on mainframe computers available to users elsewhere through terminals, modems, and phone lines. Timesharing was a huge business in the 1970s because it gave people access to computing power that was available in no other way. Much of the money that DRI made came from connect charges for corporate planners who stayed hooked to DRI's mainframes for hours of analysis. It was a successful, high-margin business focused on large corporations, financial institutions, and government agencies.

The business changed overnight in the 1980s. Personal computers on analysts' desks gave them the computational ability they sought and made timesharing obsolete. Economic data became a commodity item as low-cost competitors sprang up. "Boutique" competitors offered simpler models, cheaper data, and better contracts than DRI.

As DRI's timesharing business withered, the company never committed to something new. It clung to a shrinking base of hard-core customers, dabbled in newsletters, offered data on diskettes, toyed with the World Wide Web, and tried an ill-fated, low-margin push into consulting. It finally phased out timesharing in the mid-1990s, fifteen years after the original threat and ten years after the serious damage had been done.

This sad tale of indirection and indecision raises a question: How can leaders maintain leadership?

Proven Strategies for Holding the Lead

Three snares lurk at foot level to trip up technology leaders. First, some leading companies simply forget how they got to be leaders. They forget that they achieved their position by discovering needs and addressing them with the exceptional value-added provided by technology. Leading companies that forget this begin to focus on technology for technology's sake. They create advances that make no money. R&D becomes a hobby, not a lifeline.

Second, wildly successful companies begin to push product and forget about providing new value. The money seems to be pouring in, so they think, "We're in business to make money, so let's sell product." Digital Equipment got caught by this snare in the mid-1980s. Its revenues grew fast—for a while—but DEC produced no consequential advances after 1985. Pushing product without offering added value is the way to lose leadership in any business, even sandpaper, but it is fatal in technology. Remember, in most performance-oriented markets where technology provides value, customers buy your stuff because it's valuable to them, not because it's cool.

Third, companies follow their products down the TAS as the profitability goes out of them. Many companies go wrong by investing more and more money to capture lower and lower returns. Many think they are sticking to their knitting, whereas, really, their knitting is sticking to them. Leadership of a technological commodity requires radically different skills from leadership of exotic or specialty applications. Commodity leaders must invest heavily in production equipment to get their costs down. The skills built up in product development and personalized sales coverage won't be nearly as useful in the commodity segment. Rather than stay with an application that is becoming a low-value-added commodity, why not stay a leader and find new fields to lead?

Compete Against Yourself

Maintaining leadership demands that you move into new areas with no one to guide you. Followers always have a ready frame of reference: the leader. The follower's strategy is simple: Replicate what you can, walk in the leader's footsteps, and try to keep up. But leaders must challenge themselves. They push themselves forward, set targets for themselves, establish aggressive goals and timetables, and pit separate teams against each other. They compete against the clock. They try to build the best solution to each problem.

In charmingly named Clover, South Carolina, the president of Huffman Corporation is a man named Roger Hayes. Huffman manufactures huge state-of-the-art grinding machines that make artificial knees, turbine blades, and high-precision engine parts. Hayes is a leader in many ways, but he really stands out in one way.

Once Huffman designs a new model, he immediately starts thinking of that brand-new machine as the competition. He seems not to do this consciously; it just happens. Even though he just led its design and construction, he's already coolly assessing its technical limitations and market weaknesses. No one knows the machine and its customers the way Hayes does, so he is a formidable competitor—especially to his own machines. He acts as his own Red Team, competing against himself, probing for weakness. Since he never stops doing this, his machines get better and better, and every customer wants several more of them. Hayes is the model of the leader competing against himself.

Do this, and no competitor behind you is likely to surprise you or pull past you. The best senior executives keep their teams in the lead by holding up a mirror to them and telling them, "Compete against this fella."

Break Your Markets Into Smaller Pieces

Leaders must strike a balance between the drive for profitability on the one hand and the costs of product development on the other. Leadership produces nice profit margins, and any leader is loath to spend more than necessary, especially in the uncertain area of bringing out new products. The tension reflects why monopolies innovate so little. Why do more if you don't have to?

But those undeveloped areas that you as the leader leave unexploited are little petri dishes of opportunity for scramble competitors and your followers. Leave these opportunities out on your doorstep for long and little things start growing in them, things like competitors. One way to act as your own competitor is to break these opportunities into separate pieces and preemptively develop new products especially for them. Deal with each of these smaller segments and niches separately, just as competitors would. Assign a manager or a small team to cover each new territory. Have them compete against your larger organization in this broken-off piece of the market. (We talk more about organizational issues in Chapter 9.)

Once you occupy these smaller niches, you make it much harder for a scramble competitor or follower to find one it can exploit. Target a niche, fill in the product holes, and assign separate staff. John Deere and Caterpillar do this extremely well. They concede turf to no one. They have a piece of equipment for everyone and every interest.

Innovate Relentlessly

It should go without saying that to stay the leader you must innovate relentlessly. You became a leader through innovating. How can you justify pausing?

The push to innovate relentlessly has both offensive and defensive aspects. Offensively, you can continue to push into new territory, and the very newness of these areas assures you of continuing leadership. Defensively, your discoveries and advances can be patented and protected from encroachment. Very few leaders who continue to innovate and bring out new products are overtaken.

Speed Up the Pace

As the leader, you have the opportunity to set the tempo of innovation, expansion, and development. In Physics 101, we learn that doing the same work faster takes more horsepower. The same is true in business. By requiring more speed from your competitors, you increase what they must spend in dollars and resources to keep up with you.

Take a glimpse at the market for flat-panel displays to see three giant players speeding up the market to wear out their competitors. Flat-panel display screens are used for laptop computers, personal videocassette players, and really big, flat-screen TVs (which are commercially available in Japan). Three major Japanese makers of flat-panel displays, Sony, Hitachi, and Fujitsu, set the tempo of development. Each is huge, each wants to prevail, each is spending heavily to be the only remaining leader. The tempo of development is furious. A new advance in technological performance—brighter display, finer detail, faster speed—used to come out every several years. Now a new development comes out every eighteen months. By the end of 1998, the cycle will be twelve months long. This acceleration seriously handicaps new competitors.

Swallow Up Smaller Companies

Leaders, especially those committed to maintaining their leadership, gleefully grab whatever technological resources enable them to add value. Sometimes, rather than buying or licensing the technology, they buy the entire company. This technique is well established in the nontechnology world but seen less frequently in the technology sector. There should be more of it.

Silicon Graphics, which makes big graphics workstations, wanted an advanced animation technology to round out its product line. Instead of trying to write the software itself from scratch, Silicon Graphics went out and bought Alias Research of Toronto, Ontario. Alias made the fabulous distortion-producing software for films like *The Mask* as well as for great TV graphics and animated logos and visuals.

RPM, a specialty coatings manufacturer, has for years quietly bought companies in its area of specialty paints, sealants, and coatings. RPM's acquirees have a technology connection to the parent. RPM uses its in-house technologies in manufacturing operations, regulatory compliance, formulations, and testing to spread ideas across their huge line of products. Having grown to $1.3 billion by acquisitions and internal growth, RPM remains on the prowl, looking for little specialty outfits such as Floquil Paints—and bigger ones such as Tremco Sealants—to plug into their mix.

None of these companies practice a policy of NIH ("not invented here"), which would ensure inevitable loss of leadership. The basic formula for using acquisitions to maintain leadership is to identify niches, acquire to fill in, keep up with specialized advances, get the best technological positions, get the best people (they're *always* in short supply), get the key market space, and dominate.

These smaller businesses might eventually become serious competitors down the road, so why not apply their skills to your advantage instead? Besides, they may have a number of operating methods that would benefit your company. Here's an example in reverse. In the late 1980s, by virtue of an earlier acquisition, John Wiley and Sons, a mid-size New York publisher, owned a small outfit in the Philadelphia area that published medical refresher texts for subjects taught in medical school, a classy Cliffnotes for medical students. The little company used a number of operating shortcuts to bring out new books quickly and cheaply without reducing quality. The little publisher's finances were classic TechnoLeverage numbers: fast asset turns, high margins, great sales per employee, strong growth. Wiley's numbers were abysmal: low return on assets, low asset turns, extremely low sales and profits per employee. Wiley could have used this little company as an operating template for improving its overall corporate performance. Instead, it sold the little publisher despite having in it a recipe for improvement worth millions for Wiley's shareholders—because its line of books was not compatible with Wiley's publishing strategy. Seven years later, at great expense, Wiley put the same improvements into effect. Wiley could have done so years sooner by learning from its small acquisition.

A Few Thoughts on Followership

To my mind, followership in technology has little value. The few businesses I have seen try it were disappointed. Strategic followers give up all hope of obtaining the lush margins that come at the front end of the TAS. In exchange, they're assured of being a me-too bidder everywhere they compete. Their sole advantage is in selling to the purchasing department, where they promise, "We're just as good, and cheaper." Yet they still have the expense of achieving parity with the leader. Further, they *always* sell to purchasing, never to engineering. There is just not enough profit in the technology leader's slipstream to pursue this strategy successfully. Look at AMD following Intel for years—repeated promises of profitability, but always laboring to catch up.

Nashua Corporation in Nashua, New Hampshire, officially tried the followership strategy for a number of years in labels, diskettes, converted products, reconditioned toner cartridges, copy paper, and other products. Nashua would let others innovate, and then it would duplicate the advance. But even that took time and money. Despite having capable engineering and scientific talent, Nashua never got enough profitability out of its "new" areas to pay itself back. Nashua sold off pieces of itself to stay afloat, cut back on research, and steadily shrank until it was down to the core York photo-finishing operation. There, at last, it was—ta-dah!—a leader in efficiency, innovation, operations, and quality. By that time, however, photo finishing was under assault from electronic imaging. No rest for the TechnoWeary.

Generic drug manufacturers are determined followers. They see that Company X has a great product and patent protection for seventeen years, and so at seventeen years and one day the generic manufacturer has its product on the market. But I wonder if they are really strategic followers or just parasites living in the shadow of difficult FDA approvals. No generic drugmaker I know of has used this strategy outside of pharmaceuticals.

The penalties of followership seem to exclude it from being a legitimate technology strategy. Once parity products are widely available, the purchasing manager or consumer has one question for the follower: "How much cheaper can you give it to me than the other guy will?" The purchaser is not looking to you for any technological value-added. The question illustrates the penalty of followership. The customer now has the power of choice—competitive bids—and uses that power to drive down your margins.

It seems that for the cost required to be a follower, one can become a leader somewhere else, and more profitably, too.

A Lesson From Wingwalking

To take the lead in technology, one must constantly offer new and unfamiliar solutions. It might be useful to think of holding on to the lead in terms of wingwalking, the old county fair entertainment in which a man or woman steps from the wing of one light airplane to that of another. At some point in their act, wingwalkers must let go of the original position they have been hanging on to.

This must be difficult the first time. They have a new position to go to, something new to step onto and something new to grasp. For a split second, they need to have two handholds. Then, in less than the blink of an eye, once they judge that the handhold on the new wing is secure, they must make their step to the new wing. They must do this repeatedly, in front of big audiences, all wondering if they'll fall.

To stay the leader, you must repeatedly make your move to new markets. You have to keep repeating the cycle. Wingwalking and market leadership are not for the timorous. One excels at wingwalking or achieving leadership by doing it repeatedly. You learn to concentrate, stay calm, prepare properly, and do it precisely. There's no other way to maintain the lead once you have grasped it.

Review and Preview

As we've seen in this chapter, leaders in an application, a technology, or a market enjoy tremendous advantages. They get to set the standard, sometimes literally, and they occupy a position that others find difficult to attack. They grab the reference customers that help them win even more sales. The early, very high margins go to the leader, as does the experience that ensures continuing dominance.

To take the lead, your company must use a technological advance, a new and valuable application, to target a high-growth market. The new market need not be large. You do need a well-founded plan if you are to achieve lasting leadership, as opposed to flash-in-the-pan prominence. Once you have such a plan, then you must get started, move quickly, and fully exploit the opportunity. Leadership can be ephemeral, as the fifth TechnoLeverage Tip implies: *What one engineer can create, another can copy, so keep innovating*. Leave no gaps for followers to exploit.

Taking and holding the lead is tremendously challenging. It is also extremely rewarding. For many companies that may not have thought leadership possible, with TechnoLeverage it *is* possible.

Chapter 6 opens up the vista by revealing numerous ways of achieving TechnoLeverage. In the next chapter, we turn to specific tactics for lifting the performance of your business to new heights. These are action items you can put on your "To Do" list to enable your business to fully reap the rewards of its technology.

CHAPTER 6

Tactical Technology

Taking TechoLeverage in All Directions

This chapter is about breaking out of whatever limitations have constrained your company's growth in the past. We cover seven ways of finding growth, improving profitability, and expanding opportunities for your business. Some of these tactics are so broad and challenging that every person in the company can contribute ideas to the mix. Others are narrower and require detailed legal or technical work that involves just a few people. But these seven tactics are practical methods any company can use to achieve growth. No matter how good a strategy a company has, its profitable growth depends on carrying out its tactics smoothly. Tactics put strategy into action.

Growth doesn't come easily. For example, the U.S. economy grows at around 3 percent per year, as measured by the Gross Domestic Product adjusted for inflation. Nonetheless, many senior managers set corporate growth goals at 15 percent annual gain in sales and profits. Growing a company by 15 percent annually in an economy growing at 3 percent requires hard tactical work by every single person in the company.

How can you consistently grow at five times the rate for the economy? What if you've saturated the market? What can a manager of a division with high market share do? How can you hit your ambitious targets and satisfy shareholders—year in, year out?

The answer to these questions forms TechnoLeverage Tip Number Six:

> *TIP NUMBER SIX:* **Use technology to add value, and then go everywhere with it.**

Apply your technology to new problems. Get your company out of its rut. Improve your capabilities and take them everywhere. Find people who value them. Find growth markets and enter them with improved goods and services that increase your buyers' productivity. Repeatedly develop new offerings that are more valuable than what the rest of the economy is using. These actions form the heart of the tactics that put your high-growth, high-margin plan into effect.

The old growth models—market share struggles, price-cutting, more promotion—simply won't work for more than a few in our new, technology-driven economy. They barely get you 3 percent growth. But by using technology as a managerial tool for expansion, any business—*every* business—can achieve high and persistent growth in profits and sales.

TechnoLeverage Tactics

Here are seven action-oriented, practical, manageable tactics that you can use to take your technology in all profitable directions:

1. *Seek* new markets and new customers.
2. *Devise* new applications and invent new products.
3. *Offer* whole systems.
4. *Work* with multiple sales channels.
5. *Establish* joint ventures and alliances.
6. *Create* licensing arrangements.
7. *Employ* new production technologies.

Figure 6.1: Technology Expansion Tactics

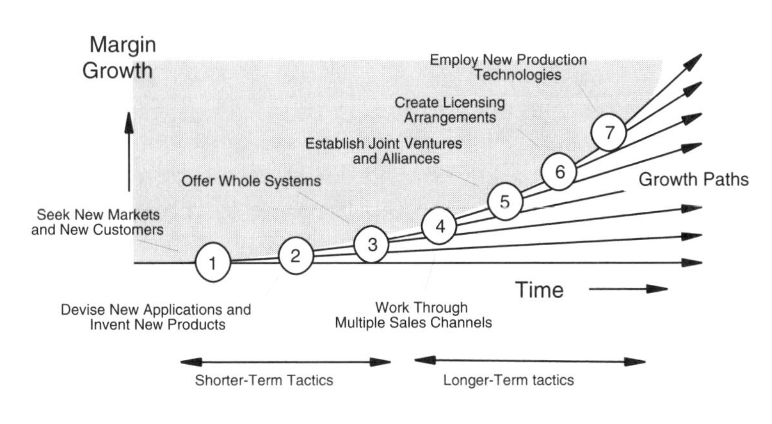

Each of these tactics has the potential to wring more profitability from a given technology or application. "Use technology to add value, and then go everywhere with it" means use as many of these tactics as you can profitably. If one tactic doesn't produce more profitability, disregard it and move on to the next one (Figure 6.1). But try them. Most companies that fall short of their 15 percent growth goals employ too narrow a range of tactics and thereby fail to realize the full value inherent in their technologies.

TechnoLeverage Growth Tactics One by One

Let's go over these tactics and show how businesses use them to improve dramatically their size, growth, and profitability. After that, let's compare two companies to see how great a difference these tactics can make.

Seek New Markets and New Customers

Start with this tactic; it has the fastest and most predictable payoff. It's the essence of the phrase "go everywhere." You and your company or division have probably been concentrating on one or two markets. Now it's time to think more broadly. Think of every new market and new customer you possibly can. Systematically take your technology or its application to each of these markets and to the attractive customers within them and evaluate the response you get.

Seeking new markets and new customers is the cheapest and surest expansion tactic you can use to attain TechnoLeverage.

In Chapter 4, on scramble competition, we described how scramble competitors seem to come out of nowhere. These are technology companies "going everywhere with their technology" in pursuit of new fields to harvest. Capturing the expansion possibilities in these new markets strongly motivates scramble competitors. Seeking new markets should be a tactic that you, too, use deliberately and consistently. Look constantly for new markets to harvest more margin from your technologies.

Implementing this tactic does not mean simply sending your sales force off in every possible direction; that disrupts your salespeople's focus. Nor should you necessarily create lots of specialized sales units, since that may be expensive and difficult to manage. Instead, the tactic should spur you to broaden your reach into specific new areas using dedicated resources, as we describe below and in Chapter 9.

One good example of exploring all possible markets is Abitibi-Price, originally a Canadian paper manufacturer and now merged with Stone Consolidated to become Abitibi Consolidated. Abitibi-Price made groundwood paper. Groundwood is produced by forcing logs against a gigantic stone some twenty feet across, spinning at high speed. The stone grinds off tiny flakes of wood, which are then finished into low-quality paper. Groundwood paper is lighter, weaker, slightly rougher, and shorter-lived than the regular grades used in better magazines, greeting cards, and photocopy paper. Newspapers are printed on groundwood; it is the soft, almost fuzzy, off-white paper also used in inexpensive writing tablets. Groundwood can be as much as 30 percent cheaper than regular paper (called kraft). That's a big difference, but most groundwood paper companies don't exploit it.

No one had ever really conducted an intensive search for new markets for groundwood paper. In four years of doing so, Abitibi-Price built up a $50 million division from zero. Abitibi considered every conceivable use for groundwood papers, even paper caskets for the funeral trade (they're used in cremations). Abitibi got out of its rut and found many new markets, including one-time-use manila folders, government payment envelopes, direct mail bounce-back cards, and even tampon inserters. Abitibi looked for any application where a less expensive, less durable, bulkier, and readily biodegradable paper could be substituted for higher-performing but higher-priced kraft paper. Relentlessly, customer by customer, new use by new use, Abitibi-Price sought out markets where the high volumes and light-service requirements justified a cheaper, slightly lower-performance material. At the end of four years, through great perseverance, Abitibi had built the largest specialty groundwood paper business in the world. It kept two giant paper-making machines busy full-time making groundwood paper for these products for new markets and new customers. (There is more detail on the specifics of this effort in Chapter 9.)

Deciding how far to persevere can be critical when seeking new markets. When taking an application to new customers, let the technology adoption-diffusion model of Figure 3.3 be your guide. After you take an application into a market and capture a good share of the innovators and early adopters, should you begin selling to the early majority, a new and bigger group of customers, or should you not? Depending upon the size, performance demands, and price resistance of the market, you simply may not be set up to sell on that scale. Abitibi had plenty of capacity, so it could. Had it not, it could have considered selling to innovators and early adopters in other new markets instead.

This raises an interesting question: Should you pursue different types of new customers in an existing market, or pursue similar customers in a new market? To the extent that you can profitably do so, *pursue both*. Again, resource availability might dictate sequencing your expansion or using the resources of others. Can Nichia, the little blue-light laser diode maker in Japan mentioned in Chapter 5, afford to pursue applications in both the CD production equipment market and the office copier market? It needs to think through where the returns are greatest and where the resistance is lightest.

This tactic involves your whole company. Everyone in your organization has the potential to think of a new market or identify a new customer. Not every suggestion will pan out, but everything possible should be done to get your entire organization to think—all the time—about new places to go. Good ideas can come from the most unexpected places. Encourage all the people in your organization to consider where else you could sell what you make.

How far to persevere with this new-market tactic depends on how these new markets affect your operating numbers—specifically gross margin and operating margin—and on other traditional measures of business attractiveness, such as the size of the new market, the number and strength of competitive new entrants, and the condition of the industry.

No company can afford to think about new markets for technology without considering international customers. If you have a successful application, the need for it extends across countries and cultures. Technological solutions appeal to customers worldwide the instant they are available. For most applications, the total market comprises every individual or organization in the world with the need addressed by that application. If you decide not to reach them with your offering, some other company will—and then use those profits to come after your domestic customers.

But don't think that just having a sales operation in Europe means real global coverage. It's surprising where you can find good prospects. There are unexpected and attractive markets open to your products, but perhaps not in the familiar places. What about Ecuador, India, and South Africa? As you plot your overseas new-market approach, think broadly. All of us in technology today are Marco Polos, pioneering our Silk Routes.

Devise New Applications and Invent New Products

A company pursuing TechnoLeverage never stops researching, planning, testing, and introducing new applications for its technologies. These new applications give rise to a phalanx of new products. This is how the tactic connects to the rest of the TechnoLeverage strategy. New markets produce ideas for new applications; new applications produce ideas for new products. Novel and valuable new products catapult you to leadership, and leadership produces better margins and faster growth.

Hewlett-Packard represents an excellent example of the nonstop, new-application mentality. H-P takes ink-jet technology into completely new applications, knowing that it is more than a fast, accurate way to apply ink to paper for printing. H-P sees it as an infinitely flexible way of precisely applying many substances.

For example, H-P's ink-jet technology, which so quickly and accurately applies liquids to surfaces, also finds application in health care for delivering finely measured dosages of medicines onto patches for absorption through the skin. H-P has also learned how to use ink-jet technology to accurately build up volumes of plastics that can then be cured or catalyzed into shaped parts. One such application involves making component parts with shapes that cannot be made by regular molding techniques. H-P uses the "ink" jet to apply layers of the material to the surface, cures them, applies the next layers, and so on. The real point is that H-P does not rest upon its dominant position in ink-jet printing technology. Instead H-P views the technology broadly, intently searching for new applications to extract from them all the value it possibly can.

Hewlett-Packard has scores of these efforts underway. Calculators, scanners, computer servers, readouts, light-emitting diodes, atomic clocks, lasers, whatever, it is constantly trying to leverage its skills. H-P manifests true genius at selling big volumes of products into new applications at high margins. Thus H-P invents high-value new products at breakneck speed and gets them to new customers fast. It is a living textbook of great methods for achieving TechnoLeverage.

H-P shows how to keep a business up high on the technology applications spectrum, staying in the more-specialized, higher-margin areas. You, too, should keep coming up with completely new applications for your company's best technologies. This tactic calls for the best effort possible from the technical and engineering personnel at your company, and from the most creative sales and marketing people.

Management stewardship quiz: Are you sure you know all the technologies you have in your possession and how they might be used? Some businesses don't. A large defense contractor we studied had thirty-six distinct core technologies spread across five divisions. One of these—a technology for battery design, evaluation, and testing—was hidden away in what might be called a hobby-business. This backwater in one of the divisions was run by a very bright and capable engineer far more interested in technology than in business. Since he didn't want ripples in his pond, he didn't request attention or resources from management. His business was small and sleepy, like a well-fed baby.

Meanwhile the company, enduring the contraction of the defense industry and struggling with 2 percent net margins, desperately needed new sources of business. Here was a remarkable new technology—protected by strong patents and led by a technological wizard—that had been commercialized but woefully underexploited. The solution was straightforward: Pair a business manager with this technologist and start a determined search for new applications. The search began with three basic questions: "What are the growth areas we could serve with our kind of batteries?" "What new applications for battery testing can we identify?" and "What battery technologies can we license to large battery manufacturers?" Three simple questions led to new applications and rapid growth for this unit.

In my experience, almost any large, multidivisional, industrial company has at least one ignored technology like this: low-hanging fruit waiting to be picked and put into service or leveraged aggressively with modest incremental investment.

Offer Whole Systems

After you've begun seeking new markets and customers and have embarked on devising new applications and inventing new products, what next? The third tactic for capturing more of the value in your application is to build it into complete systems for specific classes of customers.

Why do most makers of consumer stereo components offer entire systems? To capture more value, and to make it easier for new customers to purchase an entire operating unit.

A system incorporates all the components that the customer needs in order to get started. For example, manufacturers of bar code readers develop systems that include everything the customer needs to begin using bar coding: stationary hand-held scanners, cables, connections, communications and analytic software, label printers, data storage, networking solutions, supplies, support, and service.

The value added by a system can be significant. A system offers the convenience of one-stop shopping, freedom from worries about system performance, ensured compatibility among the components, and one phone number to call in case of problems. Customers pay handsomely for this value. Original equipment manufacturers (OEMs) build their businesses on this proposition. They add value by choosing and matching components, assembling them, readying the system for use, and providing advice and service. But your company need not manufacture every part of its system—or even do the final assembly—to offer a system.

Systems are worth considering for many companies, even if a single component of the system is the most valuable element. A system may be the only way to bring first-time customers into your market. Ask yourself and your planning team, "Could we expand our markets if we were to make whole systems and not just products?"

Even common products can be enhanced in this way. Toothpaste comes to mind. You used it this morning, you'll use it tonight, and you probably don't think about it. But Procter & Gamble and Colgate and Lever do. Think how they've changed simple toothpaste over the years. It still comes in a tube (usually), but toothpaste is becoming *a system for mouth care* rather than a simple product. Toothpaste started as an abrasive, to scrub the tooth's surfaces, combined with a flavored paste or gel to hold the abrasive ingredients. Forty years ago, abrasives and flavorings were joined by fluorides to prevent tooth decay. (Dentists' business is down, by the way. Should they have viewed toothpaste as scramble competition?) Then the manufacturers added whiteners to make your teeth pearly white. Now toothpaste also has antimicrobial compounds to prevent gum disease. Each of these is a platform for more promotion and improved positioning and higher sales, and higher prices and margins. What a value proposition has been placed before us—TechnoLeverage inspiration for all at dusk and dawn!

System sales tactics relate to the application search strategy. You may recall that Husky went from making molds to hot runners to molding machinery to pick-and-place equipment to factory design, moving step-by-step toward providing a total system. Husky had an explicit strategy of finding and overcoming barriers to the sale. The ability to break down sales barriers is an excellent reason to offer a system. In cases where potential customers would use your product "if"—if they had a place to put it, if it interfaced with their current operations, if they didn't have to go out and buy all these other pieces—there may be an opportunity to offer a system and make the sale. When you see a barrier, try to envision a profitable way to overcome it by incorporating the solution into a system.

The tactic of offering a system has many virtues: not leaving money on the table, differentiating the supplier, erecting defenses, possibly achieving leadership. X-Rite, a Michigan-based manufacturer of color-matching systems, has elevated this tactic to a strategy. Its core technology is to match colors perfectly, particularly colors of different materials such as plastic and metal, metal and cloth, cloth and vinyl, vinyl and enamel, and so on. X-Rite sells color-matching systems, but not the underlying enabling technology itself. X-Rite systems measure, match, manage, correct, and communicate color. From the standpoints of defense and leadership, it does not want anyone else's products connected to its products if it can help it.

X-Rite uses the systems tactic to drive its margins up. It has the breadth and technological excellence to provide the whole system, and nobody else does. Customers don't unbundle because they don't want to, don't need to, and probably can't.

Offering a system helps one-product companies overcome their limitations. X-Rite could have stayed a one-product company selling a laboratory color-measuring device. But it realized that the systems approach provided more growth possibilities. A number of chemical companies have begun to take a systems approach to their products. A large ingredient company we know has begun to provide its products for the personal care market premixed with other, unrelated chemicals. It adds in what the customers also use, such as emollients, neutralizers, and stabilizers. This company is offering all the customers' additives as a precision premixed system, saving the customers considerable mixing time, energy, inventory, labor, handling, control, package disposal, and purchasing expense. The systems expansion tactic represents significant value for both the technology provider and its customers.

I enthusiastically endorse new-markets and new-applications tactics for all companies, but the tactic of offering systems doesn't apply to every company. In general, it expands the scope of your operation, and it can require considerable changes in how you sell. Systems do not necessarily require high levels of investment, but they demand more change and expense than the previous two tactics. For many companies, the expense and effort are repaid when they become harder to displace as a system supplier. Users get comfortable working with systems, and they relinquish satisfactory systems very reluctantly. Consider how tightly the legions of Macintosh users in the graphics world are bound to their Apple

Computer systems. Users' reluctance to switch systems reduces your risk once you sell them a technology-based system.

Work With Multiple Sales Channels

How can you get your product to market? You have numerous alternatives. Generally, new applications for technology are sold direct. The novelty of unique and exotic applications makes them difficult for unsophisticated distributor sales forces to master. They have no patience with long, involved sales efforts. They find it hard to handle their routine duties and at the same time find, inform, convert, and train users unfamiliar with a new technology. They want products that "sell themselves," and new technological products generally fail this test.

You can sell direct to end users through your own sales force, which you can choose to organize along industry lines (called vertical marketing, such as having a salesperson trained to cover telephone companies), regional lines (South Central sales coverage, for example), or structural lines (selling to original equipment manufacturers in several industries).

Direct sales enable you to learn from customers. Their feedback is an important part of early-stage product and technology management. The more technological value-added a product has, the less familiarity both supplier and customers will have with it, and therefore the more education both need. Customers need to know how it works, and how it affects their operations, capabilities, finances, and personnel. You as supplier want to know their sales objections (their reasons for not purchasing), how they structure their financial analyses, who competes or offers substitutes, and what other tactics help you (such as offering complete systems). You want to have control over these educational lessons since you'll be paying to provide (and take) them. The higher the technological value-added you offer, the more gross margin you have available to pay for multiple sales channels. Usually these high early margins easily cover the expense of a direct sales force.

Later, when you are selling to more customers, the application is more familiar and customers require less education and service. Then you can sell indirectly. Selling indirectly means going through independent sales representatives, distributors, and jobbers of various sorts, getting into catalogs, and working through agents or partners overseas. You can support these external parties' efforts with well-targeted direct mail, perhaps with your own catalog, perhaps further supplemented by inhouse or outsourced telemarketing and direct-response 800 or 888 lines. You can use OEMs and system integrators. You can set up your own DISC (domestic international sales corporation) for selling overseas. You can sell on the Internet. Your choice of channels is rich, and your selection depends on the characteristics of your product, customers, and business.

For an illustration of how broadly a product can be sold, let's look at motion control devices. These items combine various opto- and electromechanical technologies. They very precisely move, position, and stop an object. A motion control device consists of small motors; gear systems; linear actuators; position sensors; controls and feedback devices; frames, racks, and platforms; power supplies, cables and wire harnesses; and carriages. They are used in manufacturing, pharmaceutical, testing-laboratory, and high-tech environments, and they are sold through the full range of channels. Exotic ones can be sold direct. Familiar ones can be sold through many channels because they are purchased by a wide variety of users and are described so fully that customers don't need to see them before buying them. The familiar designs can be accurately specified in terms of size, power, configuration, inputs and outputs, even quality—for example, in standardized comparison measures such as MTBF (mean time between failures)—so you can use catalogs, independent sales representatives, mail order, and the Internet as sales channels. Motion control equipment suppliers (and their customers) clearly benefit by having as broad an array of channels as possible working for them.

Sometimes adding new technological value to familiar products requires you to work with a new sales channel. Let's say you make residential windows, and you sell throughout North America. You're a well-known outfit, so Home Depot and Handy Andy and all the other home-improvement chains carry your products. You can advertise your product widely because everyone knows what it is, what it does, and when they need it. No special instructions are needed for opening, closing, and cleaning regular windows. Every lumberyard knows how to sell them.

Now let's say you want your company to offer a new privacy glass that goes translucent white when a current is applied to it. Or you want to sell skylights made of a new glass that darkens electronically. These look like just the thing to get a technological edge, improve margins, differentiate yourself, and get into some new applications. But wait—these are completely new products. They are *really* expensive compared to your old products. Home Depot is not going to display $5,000 skylights; they won't be purchased by the Saturday do-it-yourselfer. A $5,000 skylight raises a thousand questions—installation, insulation, energy use, failure, warranty, service, lifetime, exact performance in heat and cold. Ace Hardware can't handle these uncertainties.

No, $5,000 skylights are an architectural specialty for real innovators—upscale, sophisticated, experienced, confident—who use professional installers. If you're going to sell these high-tech windows, you need a new sales channel to distribute them. You need to make arrangements with commercial supply distributors, traders in architectural specialty products, and the industrial and commercial supply houses. You need to talk to architects and designers so they can recommend them, and you have to use wholesalers to sell to and supply some of them. You have to set up a new channel. But this new channel doesn't compete with your old channel. No Home Depot customers are lost to this new set-up. No dealers complain. You've tapped a new market with your new product and reached it with another distribution channel. Now you can experiment with other new window products for this upscale market.

The leverage tactic of multiple sales channels offers several advantages: It exposes your technology-laden product to more prospects in new markets, customers that you might not otherwise reach. Multiple channels generally produce incremental sales. The margins on incremental sales should be healthy even if distribution costs are a little higher. New channels usually require little incremental expense beyond the commissions paid to distributors or sales reps. Any additional sales of the same products result in longer production runs and more net profits than you would otherwise receive. It all represents pure gravy, obtained with broader distribution. The extra sales for you keep that extra margin out of your competitors' hands.

Sales agents and independent sales representatives serve many lines of business. They usually work on commission and can be effective case by case if they work hard and are well-focused. They require intensive orientation to your product, service, warranty policies, competitive positioning, and so on. Independent reps are certainly not the same as dedicated, on-staff salespeople, because they have competing demands on their time and loyalty. When you consider entering a relationship with a sales rep, learn enough about his or her business to evaluate these competing factors. A long-term relationship with a rep helps tremendously. At the outset, the more clearly you and the rep understand one another's mission, role, and expectations, the better. He or she needs your time, attention, and regular, firm motivation and pressure. If your business is in Tacoma and your representative is in Tallahassee, you want to be visible from time to time to keep the rep's energy level and interest high. A good representative produces well and requires only moderate maintenance from you.

Certain kinds of businesses have special kinds of sales personnel called specification representatives. When a technological product has complicated or subtle characteristics, a spec rep, as they're called, may help. A spec rep in the paper business or the interior lighting business talks with people who specify products in their designs for their clients. These designers function as recommenders to the actual purchaser. For example, a spec rep in the paper business calls on advertising agencies. He or she helps the ad agency understand the paper's special characteristics for product literature and direct-mail campaigns so that the agency can recommend to its clients the paper made by the spec rep's company. At the direction of the ad agency, the printer who produces the job buys the paper from a wholesaler. Spec reps are usually salaried and are often younger salespeople gaining experience.

Managers often avoid setting up additional sales channels for fear of competing with their existing channels. This fear is based on the assumption that any additional sales channel the company chooses automatically conflicts with the current channels. The challenge in marketing technological goods and services is to devise effective, efficient, and profitable routes to market. Rarely does employing a new sales channel directly threaten an established base of business.

As with most business arrangements, the benefits of this multiple-channel tactic depend upon the goals and expectations of the parties and the structure of each arrangement. The issues of resource commitments, payments, rights, time, and geographical limits must be rendered explicit and acceptable to both the manufacturer and the sales channel. If you link up with someone who is good at sales and marketing, particularly when you are selling a hot application, broadening your sales channels can be extremely rewarding.

Establish Joint Ventures and Alliances

The prospect of mutual rewards and teamwork is the driving force behind joint ventures and alliances, which have exploded in popularity since the mid-1970s. Partnerships have been around for decades, of course, but the pace of technological change and the substantial resources needed to develop products and get them to market have fueled their growth.

By *joint ventures* I mean arrangements in which two companies agree to form a stand-alone business effort or enterprise separate from either company. Most commonly, a large company seeks a smaller company's technology or application, or a smaller company seeks the financial and marketing resources of a larger company.

Alliances are increasingly common. In an *alliance,* each company remains separate but agrees to work together toward some defined end.

Good joint ventures and alliances are characterized by:

- Genuinely complementary resources from each company
- Specific, concrete goals and expectations, in writing
- Clear agreement on responsibilities and reporting lines and of the "cultural borders" of each company

For many companies, particularly smaller ones strapped for financial and marketing resources, a joint venture or alliance can represent salvation. Poor arrangements, though, divide the joys and multiply the sorrows, draining time and energy from both parties in exchange for inadequate rewards. When things go wrong, one or more of these three characteristics are sure to be missing. Let's briefly examine them.

First, a partnership must be based on *complementary resources.* Combining two weaknesses in an effort to create strength is a doomed tactic. Some arrangements with this fatal flaw combine weaknesses in two complementary areas, for example weak technology and weak distribution; others actually combine two identical weaknesses, such as two poorly performing R&D departments, in the hope of finding strength. If each party contributes something strong that complements the other, the arrangement begins with a sound foundation.

Second, the best partnerships begin with *specific, concrete goals* rather than vague hopes for "exploring an area" or "developing opportunities." The goals of the venture should be expressed in terms of achieving certain levels of revenue and profit or winning a specific number of accounts or having a certain product with specific performance characteristics by a certain date or reaching some equally tangible goal. Specific business goals motivate both parties and better justify the resources allocated to the partnership. Concrete goals also serve as organizing principles for the conduct of each party. Squishy requirements leave unclear the matter of who does what.

Third, partnerships become high-maintenance, low-return propositions if they lack *agreement on responsibilities,* tasks and time frames, appropriate reporting lines, or cultural boundaries. Potential problems in these areas can be headed off if the parties spend enough time together at the outset. Get to know one another well enough that you can avoid the alliance if the chemistry is hopeless. Abandoning a prospective partnership beats breaking up after signing an agreement. The best joint venture and alliance agreements stipulate clear what-ifs in case one party tires of the effort or the venture fails to achieve its objectives.

Fiberspar, the tubular composites maker mentioned in Chapters 3 and 4, has an excellent alliance strategy. Fiberspar has top-notch technical capabilities, so it's a strong alliance candidate. It's not a natural marketer, so it seeks out market leaders with which to align in expansion markets. The marketing partner handles sales, collections, distribution, promotion, and related customer and channel responsibilities; Fiberspar develops the products and pushes their performance to the limit. Over time, it has surrounded itself with these market-specific alliances and has become very good at starting, structuring, managing, and prospering from them. If you think of Fiberspar as the hub, then each of its marketing alliances is a spoke reaching out to a market.

Alliances come in many forms:

- *Distribution or marketing arrangements,* which are among the simplest ways of expanding your company's reach
- *Manufacturing agreements,* basically a supply agreement for a limited period of exclusivity
- *Joint research and marketing arrangements,* usually involving a big company selling the smaller company's products and funding further research, with the small company both manufacturing and conducting research and product development
- *OEM arrangements,* in which the small company agrees to sell its basic materials or components to a larger company or systems supplier
- *True joint ventures,* funded with a balanced contribution from each partnership, with one contributing money or clout and the other contributing technological capability
- *Minority equity participation,* a form of funding and support provided to the smaller, innovative company so that it can continue to go about its business
- *Sole partner with equity participation,* similar to the above but with exclusivity

A successful joint venture or alliance is seldom a union of equals, but it is almost always a union of people with equal motivation to make the partnership work.

Create Licensing Arrangements

Licensing arrangements are attractive because if they work well they produce nearly pure incremental income in exchange for minimal effort. The most widely known example is the licensing of a strong brand name for labeling consumer goods in fashion, sports, and entertainment. Fashion designers started the trend, which has spread throughout the business world. Nike has licensed athlete's names and images to augment its technology-driven product improvements. In entertainment, companies such as Disney and MGM have sometimes made as much money licensing characters or images introduced in popular movies as they made on the theatrical release of the film itself.

Such is the power of a strong brand that *The New York Times* reports that in 1996 automaker Ferrari had profits of only $2 million on sales of $500 million but expected to earn $10 million just from licensing fees in 1997. Imagine: a licensing company with an auto manufacturing sideline! This underscores the amazing profitability of licensing. If you have something to license, you need incur only the expense of a licensing agent (working on commission) and an attorney in order to generate a pure income stream.

A company doesn't have to have a strong mass-market brand to make money from licensing, but it never hurts, especially if it has a proprietary technology or application. Once again, Hewlett-Packard provides an excellent example. H-P licenses ink-jet technology to companies that use it in dosage-control or parts manufacturing. Note that H-P has an interesting cross-licensing policy. It licenses its technology to any responsible party, provided that the party licenses its own technology to H-P in return. H-P is thus entitled to go into any attractive business that develops as a result of its ink-jet technology being coupled with another. H-P has the right to combine the other party's technology with its own, thereby positioning itself as a potential competitor to its own licensees. Those licensees may be less than ecstatic about this, but they are getting H-P's technology in the deal, and presumably they are free to go elsewhere.

The second major benefit of licensing as a tactic is that it represents a high-return way to "use technology to add value, and then go everywhere with it." No company, not even an H-P, has the resources to take a widely applicable technology into all possible applications. This is why licensing is so powerful; it places in your service, so to speak, the best minds in businesses and industries that can use your technology. They work to put your technology to use, and pay you for the privilege.

Sound licensing arrangements depend upon:

- A true *proprietary* technology or brand, with strong patent, trademark, or copyright protection
- A technology or brand that plays an *enabling* role, as opposed to an ancillary role, for the licensee
- An agreement structured so that it is *mutually beneficial*, precluding unapproved exploitation by the licensee and over- or underinvolvement by the licensor

The more of an enabling role a technology plays for the licensee, and the more exclusive the sourcing of the technology is, the more valuable it is to both the licensee and licensor. A clear understanding of the technology's exclusivity and the role it plays in the licensee's intended use makes indisputable the potential value of the technology in that situation.

More companies should consider the idea of licensing their technology once their products enter the late specialty and the commodity stages of the TAS. By licensing out to large market specialists, earlier-stage businesses can recover equity they have in the technology without taking their company into lower-margin zones.

Whether or not licensing works for a company depends mainly on the strengths of the technology to be licensed. If the technology performs extremely well and is adaptable to a broad array of applications, as are laser, bar code, and ink-jet technologies, then licensing can be the most effective, efficient, and profitable means of squeezing maximum value from it.

Employ New Production Technologies

The term *new production technologies* refers to those methods that go beyond reducing costs and substituting capital for labor. It means adopting a new production technology that specifically creates value and thereby leverages the company's performance—a lift that cannot otherwise be achieved. The new production technology should somehow transform either the product's design or its performance so that it does something new and valuable.

Here's an example: powdered metal. Powdered metal is a newer technology in which an aggregation of very finely ground metal powders is mixed, pressed into a mold to acquire a precise shape, and then baked at extremely high temperatures. Parts made by this process can be more durable than forged, cast, or machined parts. Beyond durability, powdered metal technology produces parts in shapes that simply cannot be made by older methods. For instance, with powdered metal technology you can make very small, high-performance, sharp-cornered ratchet gears that can't be cut by machine. Powdered metal lets you make parts out of exotic metals that don't alloy well. This kind of new production method produces leverage because it creates completely new items that add value in new ways, rather than simply making the same items more quickly or less expensively.

Production technologies are a natural subject for licensing. For the user, they represent easy access to improvements that can be paid for by the piece. Licensors more rapidly recover their investment in development. As we see in Chapter 8, there are many technologies out there that may apply to your business, so broad and constant scanning certainly repays itself once you apply these new production techniques to your business.

The main goal in improving your production technologies is to stay at the forefront. Cost reduction and faster production are certainly worthwhile goals. However, the quality, performance, and leadership characteristics that new production technologies bring enable you to make completely new things. Use a production improvement at the first opportunity you have to employ it affordably. Newness produces the greatest leverage, the highest margins, the biggest earnings boost, and the best opportunities to establish leadership.

Raising the Bar in Bar Codes

Two companies in the bar code industry provide a good side-by-side comparison of how dramatically you can improve your performance by persistently using these TechnoLeverage tactics. Bar coding comprises equipment, supplies, sensors, software, engineering, computer systems, and services. As this is written, the bar code business is growing at 20 to 30 percent annually and has been for ten years. Since bar codes have the characteristic of being able to go everywhere, a company in this business should take that technology in many directions to maximize its value. Let's give these two businesses fictitious names: Buffalo Bar Code and Spartan Specialty Materials.

Buffalo Bar Code aggressively uses these expansion tactics to maximize its value. Spartan Specialty Materials does not.

Buffalo is a good example of effective tactical expansion. It has been around for nearly twenty years and makes materials and equipment for bar coding. It gains leverage by combining energetic marketing and advanced technology. Buffalo repeatedly targets *new applications* for its technology, penetrates *new markets*, and then builds on that growth. The company has been aggressive about offering *complete bar coding systems* to various market segments. As a result, it has developed a variety of point-of-shipment and point-of-sale devices that capture data and route it into systems to produce usable information. (Buffalo does not, however, make the data-processing systems into which the data are sent.) Each step Buffalo takes produces knowledge, improved business practices, and funds for the next step.

Spartan Specialty Materials was founded over fifty years ago (though of course it was not then in the bar code business). It started in packaging and labeling and got into bar codes as a natural extension of that business. The company, now a division of a large corporation, makes materials for bar coding applications. Within bar coding, Spartan originally served an innovative customer segment. However, as the larger market for bar coding grew, the company never tried very hard to expand beyond its original segment. Spartan is a good example of the unadventurous, stay-at-home, get-stuck-to-your-knitting company that finds itself passed by or overwhelmed in a technology-driven economy.

Neither Buffalo nor Spartan invented bar codes, but both jumped in during the early days of the business. Once in the business, they pursued it tactically in different ways and, as Table 6.1 shows, with markedly different results.

Buffalo has turned itself into a driver of the industry's growth by creating new scanners and data-capturing systems. The company has achieved leadership in its key markets and reinforces this position by creating new products and new markets. Management is committed and future-oriented. Spartan has developed a dependent position in key markets and focuses on improving its operations and penetrating existing markets. Spartan's management focuses on current business.

Although these two companies are in similar businesses, they've come to very different places. Since sustained high growth and high margins are the goals, Buffalo Bar Code clearly represents the better tactics. Spartan Specialty Materials woefully underperforms in its industry. Energetic tactical execution, compounded over many years, has clearly differentiated these two companies. This is the difference that TechnoLeverage tactics can make in your business.

Table 6.1: The Impact of TechnoLeverage Tactics

Characteristic	Buffalo Bar Code	Spartan Speciality Materials
Tactical mix	Many tactics	Few tactics
Major markets served	Five	One
Position in key markets	Leader	Dependent
Product offerings	Many	Few
Product positioning	Speciality	Commodity
Segment share	High	Medium
Sells based on . . .	System performance	Price and delivery
Sales channels	Four	Two
Strategic partners	Three	One
Engineers work on . . .	New products	Production process improvements
Marketers work on . . .	Creating new markets	Penetrating existing markets
Senior managers work on . . .	Future business	Current business
R&D spending	10% of sales	Under 1% of sales
Results		
Annual sales growth	35%	2%
Net profit margin	Over 8%	Under 1%

Review and Preview

This chapter has shown how to carry out TechnoLeverage Tip Number Six: *Use technology to add value and then go everywhere with it.* We've covered specific tactics for taking your technology in all directions and described seven expansion tactics in detail:

1. Seeking new markets and new customers
2. Devising new applications and products
3. Offering whole systems
4. Working with multiple sales channels
5. Establishing joint ventures and alliances
6. Creating licensing arrangements
7. Employing new production technologies

Use as many of these tactics as you can, at the right time, to wring from your technology all of the value that resides within it. Please remember that no one company can do it all alone. Even technological giants rely heavily on joint ventures, licensing arrangements, and other tactics that extend their reach.

In Chapter 7 we turn to the financial aspects of managing technology. Certain financial indicators, interpreted differently from how they might be for a business not using technology, help you keep your technology strategy on course. We will see how much financial managers can offer their companies when in pursuit of TechnoLeverage. We said this book was about making money with technology, and the next chapter gives you the financial management tools to do so.

CHAPTER 7

At the Controls

Interpreting the Financial Feedback

Financial managers in technology companies work at a political disadvantage. Few of them have technical degrees. They've spent their academic and professional lives acquiring skills in managing and measuring the financial aspects of companies. Any technical knowledge they've picked up is usually fragmentary and secondhand. Because of their backgrounds, financial managers in technology companies are rarely seen as integral to the "real" work of the company. As a result, in a typical technology-based company, the financial manager plays a secondary role.

Financial managers need to play an active, leading role in shaping the business and guiding its technology—from a financial perspective. In this chapter, we describe how financial managers can overcome their disadvantage. We also discuss tools that a manager in any function—executive, technical, sales, service, marketing, manufacturing, legal, or administrative—can use to manage technology better. In doing so, we give the entire management team the financial measurement and decision tools it needs to manage this high-spirited quarter horse called technology.

Technology Businesses Are Hard to Run Well

In the previous chapters, we've seen how technology increases the complexity and risk of doing business. The technology itself is new and sophisticated and heading off in new directions. Managing the company to take the technology in these new directions poses substantial challenges. The company must undertake numerous unstructured tasks:

- Understanding unfamiliar markets
- Avoiding costly dead-end development projects
- Repelling new competitors and outrunning imitators
- Finding and keeping savvy employees
- Building an organizational structure that works effectively
- Setting profitable prices for customers who can choose from many alternatives

Routine business decisions are more complicated and risky when technology is involved.

In the realm of finance, technology presents plentiful opportunities for investment and spending. Financial managers confront a feeding frenzy of funding requests. Partners, acquisitions, competitors, licensors, distributors, and technological advances all present financial opportunities that can make or break the business. The lack of proven models makes it easy to overinvest or underinvest, invest too soon or too late, or err high or low in terms of technical performance. Technology also demands continuing investment in R&D, raising questions of how to balance spending between basic research and product development, how much to spend, and how to judge whether the money is well spent.

The rapid pace of change in technology increases the need to look ahead. Many managers resist doing so, claiming that planning beyond eighteen or twenty-four months is fruitless. Although the world continues to change rapidly, two- to four-year plans and contingency plans are invaluable if one keeps a sharp eye on all the factors affecting the business. Broad market understanding and sophisticated financial analytics represent the best basis on which to create a plan that helps a business prosper with technology during periods of rapid change. Without careful planning, a technology company can easily become a victim of its own success, analyzing only today's performance, overestimating its power, underestimating its competitors, trying to support a lofty stock price, and eventually getting pulled into the commodity game. To have a long-term future, you need a long-term plan.

The Financial Manager Plays a Key Role

Financial professionals must lead the company by broadening the scientists' and engineers' attention from technology to other critical subjects—such as finance. Prompted by financial managers, the company must identify good growth opportunities and select the best ones, find potential venture partners, develop effective financing strategies, price wisely, and establish its reputation with shareholders and the investment community. Simultaneously, it must keep intelligent, aggressive, independent technical and scientific people working within a framework of management controls. Doing so requires diplomacy, information, communication, and balance. The best financial managers long ago abandoned anything resembling a bean-counter posture. Good CFOs and controllers in technology businesses have the timing, agility, and poise of the best gymnasts. They are plugged into all areas of their outfit. They're not naysayers; they are enablers. They tell technologists, marketers, production personnel, engineering staff, and the rest of management, "Here's how we can do it."

To play this role, financial managers have to identify the right indicators for the company's "control panel," and they must help the rest of management interpret the feedback correctly. In this complex setting, managers need to know more than their bottom line, their market share, and their stock price. It falls to the financial manager to educate the entire team on how to run the business.

Financial managers have the ability to do this. I have found a high level of excellence among financial professionals in technology companies. The average senior financial manager achieves excellence in a technology business more readily than the average general manager, although I'm not certain why. Perhaps there is a farm system of venture capital and accounting firms supplying financial talent. Perhaps a self-selected sample of highly skilled financial people is attracted to tech companies for the stock options and IPO potential. Or maybe, as "process people" engaged in structured tasks, certain finance professionals grasp the larger picture and readily understand technology and technologists. Whatever the reason, I usually find first-rate financial managers in technology companies.

In fact, financial managers may be the most qualified people to turn around a technology company that has lost its way. A dyed-in-the-wool technologist would blanch, but it's true. Financial managers have the advantage of being dispassionate about technology.

For a good example of this aptitude, let's look at Unitrode, a Merrimack, New Hampshire, maker of integrated circuits (ICs) for power supply management. In late 1990, after five years of stock price declines and losses in four of the five previous years, Unitrode's board brought in new managers with strong financial backgrounds. They knew little or nothing about technology. This new team had no choice but to manage the company strictly by the numbers. When they signed on, the stock was down to 1½ (adjusted for subsequent splits).

For nontechnical, finance-only guys, the results since then have been exceptional. Using what I call the "arborist" financial model—shaping, pruning, transplanting, feeding, and surgery for a sick business—these finance-only managers grew shareholder value from $39 million in December 1990 to over $1 billion at the stock's peak in 1997. The price rose from 1½ to 42 in six years.

Following a process that should be the subject of financial case studies for years to come, Unitrode management did the following:

- Set return on equity as their measure of success.
- Established a time frame for recovery (five years).
- Identified benchmark companies to compare themselves against.

- Improved the internal control systems by developing a system of reporting and evaluation.
- Stopped the losses through cost cutting and withdrawal from unprofitable accounts and product lines.
- Focused on the one business—integrated circuits—that had high return-on-equity possibilities for the long term.
- Gave the IC business the funding and engineering it needed.
- Pruned away three-quarters of the company—four operating divisions with low returns—after making sure they'd be successful after detachment from Unitrode.
- Expanded manufacturing capacity for ICs.
- Bought back and dispersed a big block of stock.
- Improved the reward structure and attracted new management personnel.
- Improved the processes for product development.
- Continued to build up cash.
- Created a supply arrangement with another IC manufacturer for additional capacity should it be needed.

From a messed-up, confused organization with no perceptible value-added there emerged a sleek, profitable, technology-intensive business under the strict tutelage of nontechnologists. Table 7.1 offers a quick side-by-side comparison to show the results.

This performance makes an important statement: You don't have to be a TechnoWhiz to build—or rebuild—a technology company. Astute financial managers can do it, too. Sound financial management is vital to the performance of a technology company.

Table 7.1: Changes at Unitrode Created by Financially Oriented Managers

Measure	Unitrode FY1990 ($millions)	Unitrode FY1997 ($millions)
Sales	$124.9	$177.9
Operating margin	7.1%	30.0%
Net profit	–$2.9	$30.2
Net profit margin	(loss)	17.0%
Return on assets	(loss)	19.0%
Market capitalization	$40.4	$1,015 (peak)
Market capitalization to $1 of sales	$0.32	$5.64
Annual sales growth (last 3 years)	–3%	16%
Cash on hand	nil	$61

Source: Value Line

Finance executives can be and should be key players in technology businesses. More than anyone—maybe even more than the CEO—the CFO's job is to convert the company into a shareholder-value-generating machine. The CFO's job is to help the whole team compare and choose business opportunities for using technology, and then make internal process improvements so the company can grow and improve its performance. This is the CFO's role in TechnoLeverage. Its importance cannot be overstated.

What the Indicators Indicate

The best measures for financially managing a technology company are well-known financial statement values and ratios. They are found in most annual reports and few require any special computation. I've resisted the occasional impulse to calculate esoteric measures of "technology intensity" or to create new yardsticks for value added. There is no need to do so. The best measures of financial health for technology-based companies are the existing measures. We just have to understand how to use them.

Interpreting these measures requires careful thought. Operating in a technology environment requires placing different weights on certain measures of financial health and activity. Nontechnological businesses should apply these different weights to their technology-based activities.

Gross Margin: How Effective Are We?

From a business perspective, technology exists solely to add value. Businesses search for applications for their technological capabilities and produce products that add value. Gross margin measures the overall success of this effort. It is the basis for TechnoLeverage Tip Number Seven:

> ▶ *TIP NUMBER SEVEN:* **Gross margin is the best measure of the value added by a company's technology.**

The higher your gross margin—calculated by subtracting cost of goods sold from sales and then dividing the result by sales—the more relative value you're adding. High gross margins, particularly 50 percent or higher, indicate that customers are willing to pay significantly more for what you make than what it cost you to make it. That differential is your dollar-denominated value-added. The TechnoLeverage strategy employs technology to add value. Companies using TechnoLeverage are heat-seeking missiles aimed squarely at gross-margin opportunities.

Gross margin represents a pure measure of performance. It is difficult to distort because its components are negotiated with third parties. The sales figure results from the interaction of a company with its customers; real customers have agreed to a sales price. The cost figure results from the interaction of a company with its suppliers and employees; real suppliers and employees supply the material and labor at specific costs. Management combines these inputs to get maximum value from them. These purchases and sales are made through market transactions by and large untainted by accounting tricks. They are difficult numbers to pump up or knock down. Gross margin is therefore a highly reliable financial indicator.

It's easy to lose sight of this utility. A client of ours had targeted 30 to 40 percent gross margins but seriously considered an undertaking offering 20 percent. That's a bad bet with little room on the upside, and upside is nice to have given the relentless downward pressure on gross margins. Some technology companies consistently achieve gross margins over 65 percent, some as high as 80 to 90 percent. They get those margins because they target them and then shun lower-margin opportunities.

Targeting 50, 65, or 80 percent gross margins may strike you as overly ambitious or unrealistic. But some companies get them. It's worth the effort to seek them because if even a small sector of your business achieves these levels of gross margin, that small unit can substantially boost overall returns, particularly if its market is growing.

Trend in Gross Margin: Are We Maintaining Our Value-Added?

Tracking the trend in gross margin on the technology applications spectrum gauges the decline in value-added of an application as it moves from unique to exotic to specialty to commodity. As the gross margin of an application declines—and over time it always does—it signals that the company must develop new, high-value applications. These new applications with higher gross margins on the TAS improve the company's overall performance.

At the same time it is renewing its gross margins with new applications, the business must squeeze more value out of its existing applications. Without these counterbalancing efforts, it slides down the TAS gross-margin curve into the low-margin commodity zone. For the company to simply continue what it has been doing without improving its position almost guarantees failure. There is no autopilot function on a company's financial dashboard.

The pursuit of high gross margin leads you in the right direction: toward higher-value new applications and away from older, lower-value ones. Here's more good news: Gross margin can be easily estimated for products in development. You can learn how much innovator customers will pay by directly inquiring, and by preorder solicitation. You can learn your costs by getting initial estimates from suppliers. By contrast, the widely used return-on-investment numbers usually applied to new-product efforts are unreliable in a technology business. They're flawed by inherent assumptions about unit volume and proposed levels of investment.

Many managers in technology companies assume that they can make up for erosion in gross margin by improving efficiency in selling, general, and administrative expenses (SG&A). Maybe they can, briefly. At best it's a short-term fix. Cost cutting in SG&A expenses as a tactic to offset declining gross margins rarely creates growth or adds value for the customer. Declining gross margin means that in the customers' eyes, what you are selling them isn't adding much value anymore.

Operating Margin: How Efficient Are We?

If gross margin measures how effectively the company adds value, the operating margin measures how efficiently it delivers value to the customer. Operating (SG&A) expenses are all necessary in bringing an application to market. The lower these costs, the better, as long as the company spends enough to enable the outfit to find, support, and retain satisfied customers.

A company with high gross margins and low operating margins excels at creating value but flounders at delivering it. This often occurs when great technologists turn out to be poor businesspeople. Their companies can achieve high gross margins and terrible operating margins. Potential causes for this disparity include poorly directed and managed sales forces, ineffective marketing, costly sales channels, expensive R&D efforts, heavy administrative staff, lavish offices, and simple inattentiveness. Pairing good businesspeople with technologists is the best way to prevent or overcome this particular failing.

Operating margin, calculated by subtracting operating expenses from gross margin and dividing the result by sales, represents the total return from running the business, before capital costs and taxes. Operating margin is critical to watch because it provides the cash necessary to finance the business. To finance itself out of its operations (which should be the goal), a company absolutely requires high operating margins. If operating profits are low, a growing business must seek equity financing (diluting shareholders' ownership), borrow (with interest charges reducing net income), or grow more slowly (undercutting its leadership position or sales reach). High operating profits are critical to growth. A growing technology enterprise with a low operating margin is a potential train wreck.

If the company maintains a healthy gross margin while the operating margin heads south, management must scrutinize the cost and performance of its sales force and that of its engineering and administrative staffs. While they're at it, management should examine its own efficiency and salaries. The latter should have a large incentive component and significant long-term compensation in the form of stock options so that management's interests are aligned with those of the shareholders.

Here's another suggestion for helping top management of medium- and large-sized businesses track the SG&A operating performance across several operating groups. Divide the operating margin of one group by the gross margin of that group; then do so for the other groups. This calculation gives you a comparable measure of how well each "overhead" group delivers value. You have a relative measure of how efficiently these functions convert the gross margin they receive from manufacturing into operating margin.

These three margin measures—gross margin, trend in gross margin, and operating margin—are vital measures for managing a technology company or a technology-focused division of a large company. All parts of the company contribute. In simplified terms, technologists and production people work on the gross margin. R&D and new-product development fight the inevitable downward trend in gross margin. Sales and administration work on the operating margin. They all work together, but they each get separate report cards.

Net Profit Margin: What Matters Is How Much You Keep

The importance of net income to technology companies is the same as it is for nontechnology companies. Net profit is the numerator for several subsequent measures, and it is an ideal comparator among technology companies. But since it holds no *special* significance, let's just mention it and move on.

Market and Revenue Growth: Catch—or Create—a Rising Tide

Everyone loves a growth market—and that can cause problems. Competitors swarm in, customers shift unexpectedly, capacity fluctuates, prices and margins firm and soften unexpectedly. But growth markets do hold major strategic advantages. Success in these markets, like kayaking in fast-moving waters, depends upon skillful analysis, approach, and execution. One can't simply pick a market by saying, "Sales of Granny Smith apples are growing twice as fast as sales of Golden Delicious. Let's sell Granny Smiths!"

The growth rate of a technology market indicates expansion of customers' need for the value added by that application. The market's need is the engine of growth for all the participants in that arena. If the application is rich enough—if it has the utility and complexity of, say, microprocessors or anti-inflammatory drugs—it requires a huge cast of participants of all types to serve the need. We've pointed out the variety of technologies absorbed in making semiconducting devices, including filtration systems, positioning systems, air showers, ceiling tiles, and high-purity chemicals, among others. Anyone who can add value to a high-growth market gets sucked in as if by a vacuum. However, decisions about which technology markets to pursue aren't as simple as those about apples. After you identify a growth market, ask the question, "How can we add value here?"

Pose exactly the same question of value in a slowly growing market, and you may get an answer that is even more interesting. NicSand, in the sandpaper business, achieves 20 percent growth in a 2 percent growth market. Interestingly, companies like our pseudonymous bar code participant Spartan Specialty Materials in Chapter 6 achieve 2 percent revenue growth in 20 percent growth markets. The difference between NicSand's success and Spartan's failure in adding value is their aggressiveness about doing so. For any company, revenue growth measures the combined appeal of its product, its approach to the market, and its sales and service, whether the business is sandpaper or bar codes.

Asset Turns: Velocity Is Critical

Financing—the right-hand, liabilities-and-equity side of the balance sheet—exists to support assets. Assets, on the left-hand side of the balance sheet, exist to support sales. So the level of sales you can support with a given level of assets is critical. The rapid revenue growth sought by TechnoLeverage means asset turnover represents a key indicator of management skill. Given the often short and uncertain life of many technology products, shouldn't the assets to support them be kept as small as possible? Consider such diminution a form of management discipline.

Asset turnover—calculated by dividing a company's sales for a period by the average level of assets employed during that period—shows how efficiently assets are employed. All else being equal, a company that turns its assets over twice as fast ties up half as much money in liabilities and shareholder equity. It also needs half as much capital to expand.

Table 7.2 shows two same-sized companies with the same attractive profitability and same growth prospects but sharply different asset-deployment strategies. Company A works hard to keep its asset base small, while Company B has a more typical approach. Company A is by far the better-run outfit from the shareholders' perspective. Company B has a far greater need for outside financing. Declining profits, slower growth, or dilution lie ahead for Company B's shareholders.

Asset turns, like margins, are difficult to fudge, and the trend in asset turns over time reveals valuable information. If asset turnover moves downward, you have a management that, one way or another, is using assets poorly. They may be adding too many assets too quickly—that is, investing too much or too soon—or acquiring the wrong assets or failing to generate sales commensurate with assets. They're buying harder than they're selling.

Asset turnover has important implications for large corporations. Loading up a new technology enterprise with a lot of assets, anything beyond what's necessary, decreases the new enterprise's chances of success. Many big companies overload a new enterprise with large-volume plants, overhead, corporate charges, staff, and other such freight as soon as the product is commercially viable, and sometimes before. This load jeopardizes the new outfit's survival. Instead, incubation periods accepted for sales planning should be used for limiting the assets applied to the project.

Table 7.2: Sample Comparative Calculations of Asset Turnover

Measure	Company A ($millions)	Company B ($millions)
Sales	$10	$10
Average assets	$2.5	$8
Average asset turnover	4	1¼
Net profit	$1	$1
Growth supportable by net profit	40%	12%

Asset turns are controllable, and the finance department can use them to keep the production and marketing groups off spending sprees. The case for financing and expansion needs to be extremely compelling. Too often in later-stage applications, the rationale is, "Gee, margins are falling. Let's invest more money to get more production efficiencies." Production says more capacity would mean more volume and improved margins. The sales force says price pressure has increased and they could sell more if prices were reduced. So the CEO OKs the investment and margins continue downward. Why? Because loading up on assets to pursue a technology-based business with declining margins is a dangerous game, winnable by very few companies that try it. Those few winners are experts at the high-volume, low-margin commodity game. The others get stuck with assets they soon wish they did not have.

Customers want value, not assets. Stop spending on assets, get creative technologically, run more value through your factory, move inventory faster, and sell harder. It helps to read clearly the feedback you get from asset turnover. Unfortunately, too many financial and operating managers fall asleep at the word *turnover*. In their slumber, they miss a root cause of financial success or failure with technology.

Debt to Sales: How Well Can You Finance Yourself?

Companies that achieve TechnoLeverage are self-financing, so they need little debt on the balance sheet. Many analysts view debt in terms of debt-to-worth ratios and measures of the ability to service debt, the coverage ratios. These approaches underscore the fact that a certain amount of debt and financial leverage is generally considered healthy. The implied logic goes: A business runs on money, and it's easier to get money from a bank than from customers, so grab as much as you can handle. But a different logic applies to technology companies.

Once the shareholders have made their investment, a technology company should strive first to finance its growth out of internally generated funds. Then, if more investment funds are needed for capital expenditures and product development, shouldn't existing or new shareholders be willing to supply the equity? I don't abhor debt, but companies that achieve TechnoLeverage need surprisingly little of it. Most of the debt that larger businesses have seems to be associated with general-purpose, long-lived assets such as buildings and production facilities, where they function as their own landlord. Very modest debt, equal to a few weeks' sales, should be the goal for most smaller TechnoLeverage companies.

Why would I believe in self-financing, given the long and successful tradition of debt financing in U.S. business? Debt financing certainly fits better with our counterproductive tax laws. Doesn't the zero-debt approach fly in the face of the time-honored practice of financial leverage? The answer is that pursuit of financial leverage entails taking on debt, usually long-term debt, as long as its cost is below the return the company can earn by investing the proceeds in the business. The net return on this debt is added to the return on the shareholders' equity. For a given level of investment, the shareholders get higher earnings, produced by the assets financed with that debt.

One might ask why this extra debt isn't useful when pursuing TechnoLeverage. Debt doesn't help because financial leverage is not uniformly useful in technology strategy. Here's why: TechnoLeverage doesn't necessarily aim for larger scale, but rather for higher margins at lower risk. Financial leverage implicitly aims for larger scale and is extremely useful in achieving operating leverage. Operating leverage calls for replacing variable costs with fixed assets in order to boost operating income with higher volumes. Financial leverage calls for financing those fixed assets with long-term debt. Long-term debt thus becomes a permanent fixed cost supporting a larger base of fixed assets. (The use of debt in takeovers, real estate, and financial services, where debt is intrinsic to the activity, lies outside this discussion.) Long-term debt is of course useful in financing fixed assets. If you are going to undertake the effort of acquiring long-term debt and long-term assets, go ahead, but do it large-scale.

The fact remains that large-scale fixed assets and related debt defeat the purpose of TechnoLeverage, which is to nimbly apply technology to create value and earn high returns. For most companies, technology-based strategy resists movement toward low-margin, high-volume commodity production, where large-scale assets are necessary.

Instead, the goal of TechnoLeverage is to go back up the TAS into the high-margin, high-value zones, using shorter-lived assets. This effort characterizes innovative organizations. A company that can repeatedly innovate seldom needs much long-term debt. Loads of debt and lots of fixed assets are too much baggage for a fast-moving technology business.

Financial leverage magnifies both earnings and losses. Given the historical riskiness and uncertainty of technology, most technology companies do well to avoid debt. All too often, the effect is more risk. If I borrow $500,000 from you, we each have more risk: I've incurred higher expenses and a repayment obligation; you've assumed credit risk and reduced your cash. We both get less sleep. We're supposed to be thinking about how to apply technology, not repay debt.

Aren't customers the proper source of "financing" for a technology business? Yes, but to get up and running many businesses require "lump assets," as I call them. You have to drill the well before you pump oil from it. You have to write the software program before you can sell copies. You cannot manufacture paper until you've built a paper mill. Lump assets don't produce until they're complete.

In such cases, up-front funding should be equity to the extent possible. When debt is the initial funding, retire it as quickly as possible. TechnoLeverage is, after all, a way for companies of all sizes, in technology or not, to prosper by developing high-value, high-margin business. Once on that path, debt becomes less necessary.

Revenue and Profit Per Employee: Everyone Counts

For several years in the 1980s, a major U.S. book publisher had total profits of about $9 million annually, and about three thousand employees during each of those years. That translates to annual profit per employee of $3,000. Think about it. For that $3,000, each of those three thousand people worked 250 days; got up each morning; dressed for work; commuted to the office by train, bus, or car; attended meetings; made and took phone calls; edited and promoted and sold and then commuted home; picked up the dry cleaning; and prepared to do it all again the next day—for what? To put about $12 on the bottom line per day. It hardly seems worthwhile, does it?

Companies seeking extreme performance keep their head count small relative to sales and profits. They seek talent, creativity, energy, commitment—qualities that create value—rather than lots of people. They enhance their staffs' accomplishments with as many productivity tools as they can. They keep their "asset base" of people small so that its asset "turnover" is high. If you keep the base small, you can get high value out of your assets, human and inanimate.

Like asset-heavy businesses, high-head-count-to-profit organizations are flabby. They generate waste through the law of large numbers. As they get bigger, they drift toward the norm and become more ordinary. Management's attention becomes too diffuse to produce extraordinary results, so it stops demanding them. Talented, creative, energetic, committed people need first-rate management to produce great results. If they don't get it, they go elsewhere.

Revenue and profit per employee are the signals to read when measuring the productivity of human assets. A downward trend means trouble. The cure is not as simple as one might conclude from the downsizing and outsourcing waves of the late 1980s and early 1990s. Downsizing and cost cutting bolster earnings in the short run, period. Most companies would do better if they increased the value added by their human resources.

One way to do this is to judiciously add assets to augment labor. The publisher mentioned above was using labor-intensive processes to produce books. Too many people, with the wrong skills, were using inadequate equipment and outmoded processes. With minor reductions in head count, more technology, updated processes, and newly attentive management, the publisher was able to increase its revenue and earnings per employee tenfold over several years.

Market-Capitalization-to-Sales Ratio Relative to Peer Group: What Are We Worth?

One raging question looms over all other financial issues in technology companies: "How much is this company worth?" Is share price times the shares outstanding a fair, reasonable, or worthwhile mark to manage to? If you're a private company or a division of a larger company, who can say what you're worth? Maybe it's whatever the banks would lend a strategic buyer, or maybe the value of the last IPO of a company like yours. It's a mystery.

More questions: Is a highly valued, fast-growth enterprise overvalued? What is the value of its managers' ability to call forth growth from many directions over many years? What is the value of its skill, focus, inventiveness, and teamwork? What is the value of being embedded in the customers' operations, activities, and expectations? Could it be *undervalued*? How should we go about establishing a correct value?

If you base your corporate valuation on earnings per share, you get an other-worldly number that fluctuates wildly. Earnings are the bottom number of the entire income statement, and every fudge factor has already slipped in. When a company valued at high multiples of earnings hiccups or experiences an order snafu, its computed value skyrockets or nosedives. Yet everyone knows the value of the company didn't really change that much. So I don't include earnings in calculations of value (though I certainly look at earnings when evaluating the company overall).

To replace earnings as a basis for measuring worth, you need to begin with a number that's hard to fake, one that factors in how well the company deals with its environment. You need to work with low multiples to reduce fluctuations. You need a number that signals what to do for improvement, a number that's easy to calculate and understand and that applies broadly. I'll go with *revenue* as a basis for valuation any day.

Revenue is affected by customers and competitors. It is available for all public and many private companies. It fluctuates relatively little. Relate revenue to market capitalization, and then use a peer group of eight to twelve companies for comparison, and you have an excellent measure of your company's value. Market-cap-to-sales comparisons reveal whether your business is growing in value, how it is valued, how it is perceived, what rewards lie ahead if you improve, what the peer group does better that you could emulate, and how much you should spend to improve the situation. (We examine this calculation in Chapter 10.)

For the peer group, pick companies that are in a similar situation from a TechnoLeverage standpoint. Try to pick peers in other industries, peers of similar size, that add value in similar ways, lead in similar degrees, and grow at similar rates. Wall Street has already picked up on TechnoLeverage, even for companies with modest amounts of technology in their mix. The market places great value on high growth with high margins and visible use of technology to add value.

Effective technology strategy increases equity valuation. Whereas an ordinary company might fetch an equity value of $0.80 to a $1.50 per dollar of revenue, a technology strategy company can produce $3.00, $4.00, or even $5.00 in equity value per revenue dollar. As an investor, why not pay extra for an effective strategy? It increases the likelihood of continued gains in revenue, earnings, and value.

This is where you get to see the real financial gears and levers of TechnoLeverage: in the equity value placed by the capital markets on revenues for technology companies. Financial managers have exceptional reasons for attending closely to this particular value-generating capability of technology. This compact combination of revenue, multiples, growth, margins, and peer comparison helps financial managers lead the way toward creating enormous value.

Percentage of Sales and Profits From New Products: What's New?

We've talked at great length about the need to innovate constantly and introduce new products for new markets repeatedly. Here's the measure that covers it: the percentage of the company's sales and profits attributable to products introduced in the last five years. This is a wonderful internal measure of vitality. Relatively few companies publish this number. If they do, you usually find it mentioned in passing in the chairman's letter or the management discussion and analysis section of the annual report.

This number is a great predictor of future success and growth. A business like 3M aims to generate 25 percent of its sales from products developed in the last five years. This number is one every CEO and CFO should know. Low sales from new products means ineffective development, weak marketing, or no focus on growth. Good sales but poor profits means me-too products, excessive sales costs, a "buy growth" strategy, or other form of weak execution. High numbers for both mean aggressive, forward-focused management.

Businesses that intend to grow profitably target these two measures explicitly, and they achieve them. Financial managers need to prod the cautious, monitor the exuberant, and unload the inactive.

Deep Mystery: The Return on R&D

Managing research-and-development funds stands among the great mysteries in technology businesses. How much should be spent? On what should it be spent? How should the work be managed? What is the return? At the macro level, companies and industries in the United States are periodically accused of not spending enough on R&D. At the micro level, almost any company can well question the return on its R&D investment. For a given company, the return on total R&D is almost impossible to measure.

Financial managers can make a real contribution to the management of R&D. I don't mean by cutting it, but by straightening out the purpose and accounting treatment of R&D, and by energetically measuring the return from it. Managerial accounting for R&D funds starts with breaking them down into tactical R&D and strategic R&D spending.

Tactical R&D aims for improvements to existing products, for product enhancements, for solving customer and production problems. These tactical efforts include making existing products smaller, faster, cleaner, or better in some other way. Money for tactical R&D should come from the funds of operating units or divisions. Tactical R&D spending can be tracked effectively using project-management software and similar tools. The best measure of the return on tactical R&D is incremental sales and margin improvements. Product-line profitability reports can usually capture this improvement.

Strategic R&D efforts aim for breakthroughs, for completely new, commercially viable applications or processes. Although it's not foolproof, one measure of return would be proprietary, revenue-generating products from these efforts. Money for this kind of R&D should come from corporate funds or, if necessary, a divisionally funded pool kept separate from operating funds. Although these funds should support the previous measure, sales and profits from new products, usually the development horizon is too long to generate current-period improvement in the measure.

Tactical R&D and strategic R&D should be funded separately and managed differently. Yet most companies have one pool of R&D money, which they usually treat as an administrative or overhead expense to be absorbed by "corporate" or allocated to business units based on revenue. Measurement of the return on R&D at most organizations remains quite hazy—if done at all. The fact is that, in practice and despite protests to the contrary, many companies see the R&D budget as a measure of their "commitment to R&D." Having made the commitment, they hope for the best and leave it at that. As with any outlay of money, the better the management controls are, the better the potential return will be.

The Perils of Forward Pricing

Another valuable contribution a financial manager can make to a company pursuing TechnoLeverage is sound pricing strategy. The key thing for a growing, high-margin business to avoid is forward pricing. *Forward pricing* means decreasing the price of a relatively new application of technology in order to win new customers more quickly and grow the market faster. Often this amounts to an attempt to rush the market, to make it really big too soon. This rarely works. Instead, it gives away money.

Think back to the adoption-diffusion curve. First the innovators purchase the new product; then the early adopters come on board. Then comes the early majority, followed by the late majority and, finally, the laggards. Forward pricing assumes that you can hasten adoption of the application by dropping your price.

Resist the impulse to cut your price in this situation. Despite the logic that "we'll sell more if we cut our price," in most cases a price cut does not accelerate adoption significantly. Why? The innovators and early adopters aren't very price sensitive; they are performance buyers. At the stage in which they constitute the market, they need the product for the high value it adds for them. Many of them actually like the competitive exclusivity that the high price gives them. They're buying a tool their competitors can't afford. Your high price protects their purchase. A lower price would debase that value.

By definition, early buyers see a high return from your new technology and can therefore handle its higher price, whether it's a unique, an exotic, or a specialty. Furthermore, innovators and early adopters willingly pay for novelty, sophistication, competitive advantage, and bragging rights.

Margins will never be as high as they are initially; why rush to give them away so soon? When a company pursues TechnoLeverage, financial managers must lobby for value pricing. Unlike forward pricing, value pricing maximizes margins. Value pricing expresses the business rationale, from the customer's perspective, for a high price. In promoting value pricing, finance must usually oppose ordinary pricing practices, such as "competitive" pricing and cost-plus pricing (cost plus a "reasonable" profit, whatever that is). TechnoLeverage demands value-based prices for high-value applications. It also demands willingness and ability to develop such applications repeatedly, to keep the company's margins and returns rich.

Targets for Technology Businesses

Let's look at some target figures for the measures and then at the numbers for a couple of specific businesses. The performance targets in Table 7.3 should be viewed as just that—targets—but they are achievable with technology strategy. These figures would be quite hard to hit in nontechnology businesses. Even so, they can

Table 7.3: Performance Targets for TechnoLeverage Businesses

Measure	Target
Gross margin	50% or better
Trend in gross margin	Steady or upward
Operating margin	20% or better
Net margin	10% or better
Fluctuations in net income	The fewer the better; no losses
Market growth	8% or better annually for large markets, higher for small markets
Revenue growth	15-20% or better annually
Asset turns	2 or higher annually
Debt-to-sales, weeks of revenue	6 weeks or less
Revenue per employee	$250,000 or better
Net income per employee	$25,000 or better
Market capitalization-to-sales for peer group	Varies with situation
Return on net worth	18% or better

be viewed by nontechnology companies as pointers toward what's possible and as benchmarks for efforts that employ technology.

I know some of these numbers strike some readers as ambitious or exotic, so I include in Table 7.4 the most recent annual values (as of this writing) for these numbers from the public records of two technology companies: Hewlett-Packard and Cognex, the latter a maker of machine vision systems and a "bump company" discussed in Chapter 2. Neither is the highest scorer, by any means. I've included them here as much to inspire action as to support the case for a high-value, high-margin approach.

These businesses have achieved TechnoLeverage. We know they are adding high value because they have attractive gross and operating margins. Everything else flows from that. From the technological standpoint, these companies' customers are willing to pay a premium for their products. From the financial standpoint, each company can fund its own operations and growth, without significant debt.

Table 7.4: Here's What's Possible: A Look at Two High-Performance Technology Companies

Measure	Hewlett-Packard ($millions)	Cognex ($millions)
Total revenue	$42,895	$155
Gross margin	34%	73%
Trend in gross margin	Slight decline	Flat
Operating margin	13.7%	13.5%
Net margin	7.3%	26.1%
Fluctuations in net income	Gains last 8 years	Gains last 5 years
Market growth	—	—
Revenue growth	18% annually	27.5% annually
Asset turnover	2.07 per year	2.66 per year
Cash on hand, weeks of revenue (large cash balances distort asset turns)	5.5 weeks	51 weeks
Debt-to-sales, weeks of revenue	5.5 weeks	0 weeks
Revenue per employee	$351,900	$384,406
Net income per employee	$25,587	$100,248
Market capitalization-to-sales for peer group	Depends on peer group selected	Depends on peer group selected
Return on net worth	19.3%	17%

Source: Value Line
Note: All figures fiscal 1997 actuals

Multiple Measures Make Money

The uniformly high measures of financial performance achieved by the two companies in Table 7.4 point up a lesson for everyone, no matter what their business: To get the benefits of TechnoLeverage, build a system of financial controls and incentives that focuses on *all* of these measures. Consider each one vital, and get your people to do the same. Most senior executives get people focused on one measure, or maybe two, for several years, to concentrate effort. The measure may be revenue growth, net income, new products, or costs. The thinking is that if the company hits "the number," the other performance measures will fall into place and everything will be OK.

Unfortunately, this one-number approach distorts behavior and impairs performance in other areas. For example, if revenue growth is the target, margins often suffer as the sales force "buys business" with low prices, taking on deals they would spurn if they focused on earnings. If operating income becomes the sole goal, the top line can suffer as quality and service sag. If revenue per employee becomes the focus, staff can be needlessly cut and assets added to make up for it. And so on.

Financial managers must broaden the focus in a company to *multiple measures.* They must get people to see each key measure as important. Then they must help the rest of the team understand what kind of company they are trying to build. They must place checks and balances in management control systems so that operating people cannot "game" the system. They must establish incentives that are based on interrelated performance measures that support aggressive business practices. These practices—adding value, value pricing, controlling costs, keeping asset and staff levels low, and avoiding debt—help the financial group lead the way to achieve TechnoLeverage.

Review and Preview

Sound financial management is intrinsic to sound technology strategy. Thus financial managers, and financially savvy managers within every corporate function, can make a major contribution to companies pursuing TechnoLeverage. As this chapter points out, this often means guiding technologists based on financial realities. It also entails practicing a special kind of financial management. For example, debt should play a relatively small role in the financial structure of these companies, and separate accounting for tactical and strategic R&D funds helps a company track and measure the return on these funds.

Financial managers in companies pursing TechnoLeverage should pay particular attention to certain measures of performance and financial health, which have been highlighted in this chapter. The most important measure for companies using technology is gross margin, which is the subject of TechnoLeverage Tip Number Seven: *Gross margin is the best single measure of the value added by a company's technology.*

Other key measures include:

- Operating margin, which measures effectiveness in delivering value
- Asset turnover, which should be high
- Debt-to-sales, which should be very low
- Revenue and profit per employee, which should be high
- Market-capitalization-to-sales, which can be made very high

The goals of TechnoLeverage are value creation, profitability, and growth, achieved with a relatively small staff and a well-utilized base of assets.

In Chapter 8 we switch the focus from finance to technology. There we examine ways of assessing your current technologies and skills, and we take a look at the incredible range of technologies waiting to be put to use in your business.

CHAPTER 8

A Manager's Tour of Technology

Seeing What's Out There and Knowing What It Can Do

In previous chapters, we have seen that executing a technology strategy requires you to add value to your business with technology, and then capture the value you've created. This chapter focuses on finding the technology that accomplishes this mission. Which technology can add value *for you*? Where can you find the technologies that will make money? Where should you begin?

Whether you are in the fire alarm system business or amusement park rides or food packaging or recreational vehicles or financial services or retailing or ophthalmic goods, we discuss introducing technology in some way into your products, processes, and markets—all of them. We ask which technologies can make important improvements or create completely new businesses.

Deeper down, this chapter deals with technological awareness. Awareness of technology is fundamental to TechnoLeverage—and to TechnoLeverage Tip Number Eight:

> ▶ ***TIP NUMBER EIGHT:* Awareness of many technologies greatly improves your choices of new things to do.**

I hope our tour of technology gets you thinking about the vast array of capabilities out there that could help you right here. Somewhere out there is at least one technology that can make you more money. More likely, there are several such technologies. Technological awareness removes the blinders of day-to-day business pressures.

If you're aware of the fabulous capabilities that are opening up today with such astounding speed, you can banish such thinking as "We're just not in a growth business" or "It's hard to come up with something new in this business" or "We've brainstormed repeatedly and we just don't have any ideas for new products." Knowledge of technology keeps all of us in touch with two facts of business life that are easily forgotten amid everyday pressures. First, there's a big world of new ideas out there. Second, if you find ways to use these new ideas, you can improve the profitability and growth of your business.

Let's go find those new ideas.

Mapping Your Technologies

Where better to start your tour of technology than right in your own company?

Many executives have found that mapping their corporate technologies is an enlightening way to begin a TechnoLeverage growth effort. During this process, all members of management develop an understanding of the technological strengths of their organization. They can also discover which new technologies would benefit them.

For this first step, get to know what you have, what you'd like to have, and what you lack. Then you can more easily fit new capabilities into your skill set, find gaps to fill, and build up your strengths. We start with a snapshot of the technological capabilities of your business, or the most valuable nontechnological ones.

Essentially, a business's technologies and competencies can be divided into four categories:

1. Core technologies
2. Supporting technologies
3. Technical and business competencies
4. Foundation knowledge

Figure 8.1: The Technological Capabilities of a Corporation

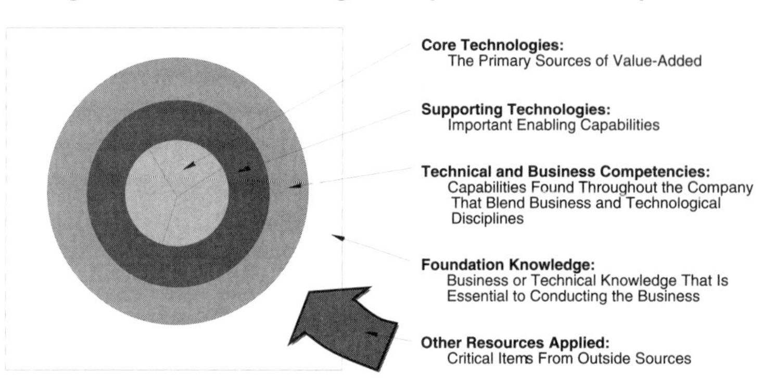

To these four we add a catch-all fifth category: other resources applied from outside. This category covers services or other intangibles that have special value-added, such as engineering consulting, industrial design, software and programming, or laboratory testing. These you regularly buy from outside and do not intend to supply for yourself.

All five of these technologies and competencies are displayed diagrammatically in Figure 8.1. The inner core contains what's most valuable to your company, while the less valuable elements are ringed around the core.

Core Technologies

In the inner ring are the *core technologies*, the basis of the company's ability to add value. They are the technological heart of the business. Given this, they need to be identified and treated as vital. Core technologies are absolutely critical to your ability to make money. A company or an industry that misunderstands its core technologies, or allows them to wither or be encroached on or overtaken, invites disaster. Correspondingly, a company that protects, enhances, and expands its core technologies sets the foundation for a profitable, high-growth future.

Core technologies should not be confused with applications, that is, with the products in which they are used. Could the decline of the minicomputer companies have resulted from their confusing their core technologies with their applications? Their core technology was designing smaller, higher-speed integrated circuits and creating operating system software that was more compact, capable, and usable than that of mainframes. Had they pushed their core design technology further, they might well have led themselves to develop the microcomputer. Instead, they thought making minicomputers (which is simply an application) was their core value-added, and they ran aground when the need for minicomputers slowed. Intel has since demonstrated that the need for improved designs for computer circuitry has never declined. Identifying a company's core technology is fundamental to technology strategy because one core technology can spawn several businesses.

To ascertain your core technology, ask yourself what you do that is most valuable. What skill of yours would you least like a competitor to have? What's the hardest to do right? What skill distinguishes your company from others? What is the hardest to copy? What do your patents cover? What would an acquirer most value about your business? This crucial value-added is vital to the company. A company must protect its core technologies and push their performance to the forefront—to leadership, as we said in Chapter 5—if it is to have a future.

Supporting Technologies

Supporting technologies are technologies needed in your business but not crucial for it to own, dominate, or lead. Relative to core technologies, these technologies are, to the company in question, secondary and sometimes generic. They can reside on staff within the company, or initially be purchased, rented, outsourced, or contributed by strategic partners, and you now use them without changing their form.

Supporting technologies are certainly important. Without them, the company cannot conduct its business properly. Yet supporting technologies are not central to its ability to create value. Usually supporting technologies are not a major focus of investment, though most companies must stay near the forefront in these areas. (Note that one company's supporting technology is often someone else's core technology.)

To identify your supporting technologies, ask yourself which essential functions you either can make yourself *or* wouldn't mind buying from others. Which of your key production or analytic skills are generic but necessary to run the business? Identify activities you conduct that all your competitors can also do, and just as capably. As an example, an engineering design firm would be lost without its sophisticated CAD-CAM design software. It can't compete—literally, it can't function—without the software. Everyone on the staff uses it to do their work. They're skilled at using it. But this software wasn't written by the firm; staff just use it to design advanced metal shapes made by various machining techniques. The capabilities embedded in the software are thus likely to be among the design firm's supporting technologies.

Technical and Business Competencies

Technical and business competencies include the full range of professional skills that the company needs in order to run the business. These skills generally reside on staff but may sometimes be outsourced, depending upon the business and needs of the company. They often are academic disciplines, types of training, or special skill sets. For instance, knowledge of site-location routines would be a business competency for part of McDonald's home office staff and for banks with extensive branch networks. Correspondingly, specific mathematical skills for site selection might be a core technology for a site-location service business. Acquisition analysis would be a business competency for anyone conducting regular searches for new companies to be added to a major conglomerate.

In many cases, technical and business competencies actually form the core of a company, enabling it to have a business without a true technology at the core. A retailer, for example, may have at its core only managerial and marketing competencies. United Parcel Service (UPS) would seem not to have a core technology. It has no scientific or engineering capabilities as the primary source of its value added. Instead it has outstanding general managerial competency. UPS's value-added arises from its managerial process, which relies on several supporting technologies such as time-and-methods analysis, ergonomics, and logistics. Using trucks, planes, fuel, labor, and dispatching and routing systems, it has created a dominant position from generic components.

Brand identity can be important in such nontechnology-based businesses, but a brand is not a core technology. Many brands, notably those of distributors and retailers such as the GAP, Eddie Bauer, and Tiffany, arise from managerial and marketing competencies rather than core technologies. Gillette is a brand-name company that relies heavily on technology—dermatological emollients, deodorants, lubricating substances, inks, plastics, metals, molding, packaging—which it uses as a platform for its brand- and product-marketing skills. Gillette's strategy of combining technology with brand marketing closely resembles Nike's strategy for its shoe business.

To ascertain your technical and business competencies, ask yourself what specialized business, scientific, or technical skills certain people joining your organization need to have. What kinds of special educational and business backgrounds do you seek? Is there a blend of business and technical competency that you look for in your top managers? What kinds of academic degrees do you need in abundance to run and grow your business? These are likely to be your technical and business competencies.

Foundation Knowledge

Finally, *foundation knowledge* represents the knowledge base the company must have to fashion its technology into applications. Specifically, this knowledge centers upon the market, customers, and end-users: their needs, the value they derive from the technology, and how they use the application. Almost invariably, this foundation knowledge must reside within the company. It's a generalized qualifier that everyone must have, a background they must share. Virtually everyone in a Website design business must have an understanding of the functioning of the Internet, of the HTML programming language and others, and of access technologies and protocols so as to participate in the business.

To ascertain the foundation knowledge of your company, look back at your recent hiring mistakes and determine what failing most quickly disqualifies these new candidates. These are the skills your managers and employees simply must have to perform the activities of your business. These crucial requirements are likely to be your foundation knowledge bases.

Mapping Technological Capabilities

Mapping technologies and capabilities in this manner enables management to see its resources more clearly. These particular resources tend to be human: the collective skills, knowledge, and experience accumulated by the organization over time. Given that they reside in people's heads, they are not accounted for on the balance sheet. It is important, therefore, to account for these resources in some way, and mapping represents one disciplined way to do so.

Here's an exercise in strategic skills inventory that several heads of R&D whom I know have found useful. After you identify your company's core technologies and skills, see which of your personnel have them. Which individuals possess your company's most valuable technological skills? Think what would happen if they were unable to work for four months or went to a competitor. Where would you be most vulnerable? Where are you most thinly staffed? Leading-edge companies of all sizes should periodically assess their skill base in this way and be sure they have backups or strong number-two individuals in key skill areas.

Case Study: "Shoreline Sensors"

Let's take a drive now, on our tour of technology, over to a hypothetical company called Shoreline Sensors. We use Shoreline to illustrate what we mean when we discuss mapping technological capabilities and then finding technologies to expand them. Shoreline makes pollution control monitors. These are a specialized family of handheld instruments for businesses, government agencies, consulting engineers, and universities. Its business has flattened out recently and Shoreline's managers would like to get into "something with growth potential," as they put it.

Shoreline's Core Technologies

Shoreline believes its core technology is designing and manufacturing electrochemical test equipment. These technological skills lie at the core of the company. Electrochemistry is a technology for electrically sensing the presence of a substance, compound, or organism diluted in a fluid medium, and communicating that presence to an observer. Electrochemical technologies apply and measure changes in electrical current caused by changes in fluid chemistry. For example, certain chemicals added to water increase or decrease its resistance to electrical current. Shoreline uses its core technology to develop and manufacture products that measure this

Figure 8.2: Shoreline's Core Technology Now

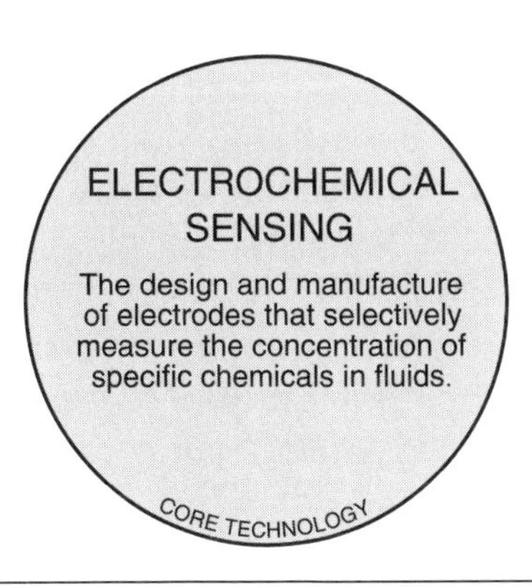

effect. Figure 8.2 illustrates Shoreline's core technology.

By examining core technology, management sees that it could consider branching out into optochemical sensing technologies because they, too, apply widely to pollution control monitoring. Optochemical technologies deal with optically measuring certain observable effects that chemicals have on a material. For instance, Shoreline's chief technologist has been fascinated with recent advances in measuring traces of water pollutants using optochemical tests that turn parts of a test strip deep blue under ultraviolet light. Understanding that optochemical sensing could enable the company to identify many more pollutants and contaminants, management realizes it could broaden the business base and add a new core technology with a move into optochemistry.

Shoreline's Supporting Technologies

Shoreline sizes up its supporting technologies, as shown in Figure 8.3. It has supporting skills where you'd expect them, in the individual elements of making its kind of instruments. From this assessment, management realizes that Shoreline has no skills in automated instrumentation routines, which it needs if it adds more tests to the roster with a new sensing technology like optochemistry. The managers also have to improve capabilities in several optical areas if the company is to succeed.

Figure 8.3: Shoreline's Supporting Technologies Now

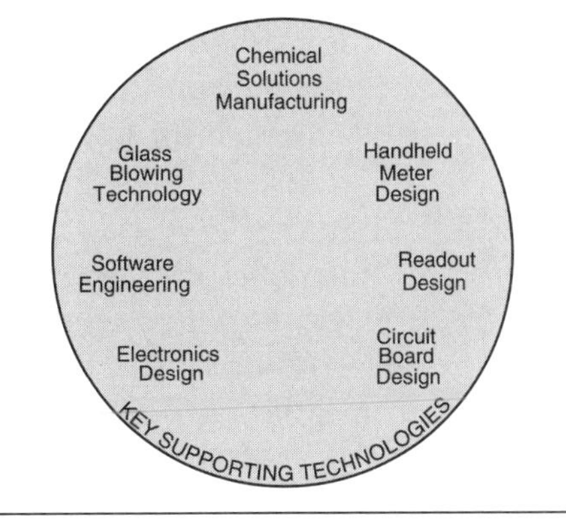

Figure 8.4: Shoreline's Technical and Business Competencies Now

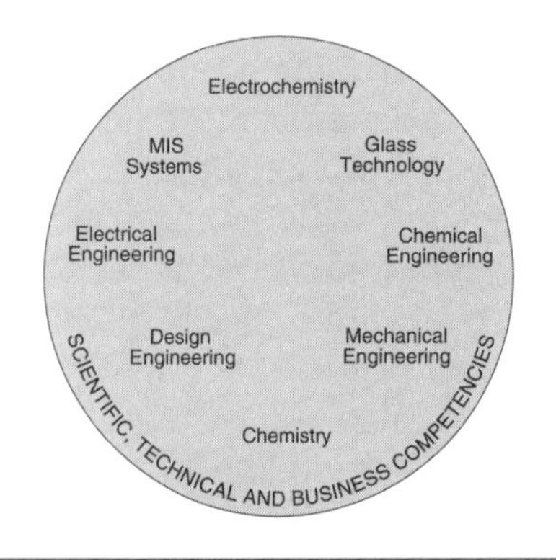

The analysis of supporting capabilities tells management that it really needs to beef up Shoreline's talent pool, and where to act.

Shoreline's Technical and Business Competencies

Next management looks at technical and business competencies; we see their analysis in Figure 8.4. Then they consider the future business environment and expect that they'll need to boost their customer service capability with more telemarketing and outbound sales. An industry magazine recently gave them a weak competitive rating in its review of the appearance of Shoreline's and its competitors' equipment. Shoreline also needs to boost the skills and size of its industrial design staff, given the new lines of equipment they have to design, build, and market.

One of Shoreline's scientists notes that the company doesn't have many plastics skills. Broader skills in this area are necessary in the immediate future because Shoreline will have many more choices of materials, and it's critical to choose the right ones because they affect the sensing processes. Also, the team sees that it has very limited optics and optoelectronics skills, two other areas that will need a big boost when they add optochemistry as a new core technology. The rest of Shoreline's technical and business competency set is perfectly aligned for the next three to five years.

Shoreline's Foundation Knowledge

Shoreline's foundation knowledge, shown in Figure 8.5, comprises mostly knowledge of field testing environments and some process testing environments. Management is aware that knowledge of lab testing is limited. This gives some executives an inkling that acquiring a lab testing business or instrument maker might be in the cards for the near future, especially if Shoreline could find one with critical skills in optochemical sensing. The CEO thinks he knows a small private company that might fit the bill. And with Shoreline's planned expansion into sales to laboratories comes a need to understand the distribution channels that serve this segment so as to supplement Shoreline's existing direct-sales capabilities.

Figure 8.5: Shoreline's Foundation Knowledge

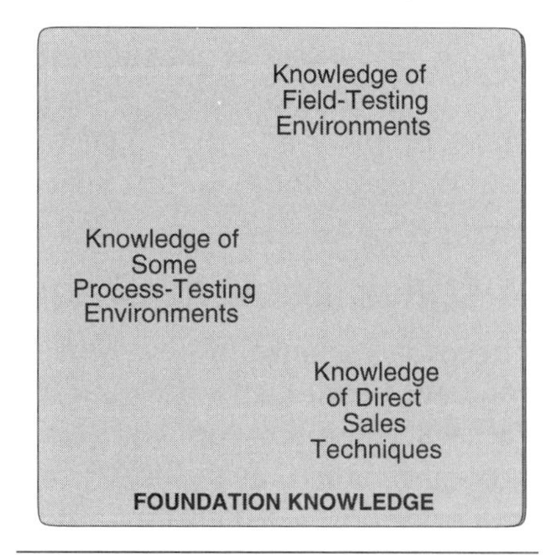

Now that Shoreline has taken a careful look at its technological self, it can plan knowledgeably for the future. It has seen the need to look outward for new technical and business skills if it is to expand into any new, fast-growth business areas like optochemical sensing.

By reviewing present structure and future plans, Shoreline's executives realize they must broaden their technological competencies. Shoreline can then use its new capabilities to develop new products, enter new markets, expand sales to current accounts, push aside a stodgy competitor, acquire a small outfit or two, and sell through more sales channels. The company sees how to grow bigger and become more profitable. It has a TechnoLeverage strategy, painted in broad brush strokes. Now Shoreline is ready to see what specific technologies are out there for it to consider adding to the mix.

An effort like Shoreline's to map technologies and capabilities can be especially useful in searching for acquisitions and joint ventures and when making decisions on hiring and outsourcing. Developing a clear picture of your company's technologies and capabilities and of their relative importance to your business enables you to judge which resources you need and their value to you. If your core technologies mesh well with those of a potential acquisition or joint venture partner, you have a compelling reason for creating a business combination that builds on these strengths. If a potential hire can contribute a critical business competency that the company lacks, perhaps the candidate's value is greater than you would otherwise think. The same is true of a small acquisition whose strategic and time-saving value may appear much larger than its small size might suggest.

So far, our manager's tour of technology has taken us to your company and then over to see Shoreline. Now let's take our tour of technology out onto the open road.

A Quick Look at Some Cool Technologies

The wide expanse of new technologies offers great opportunities for new products, high margins, and rapid growth. The list in Table 8.1 is my compilation of some of the high-potential technologies, which, as this is written, I feel present the most interesting long-term prospects for anyone deciding to use them. My subjective term *cool* encompasses underlying speed of technological development, inherent novelty, technological sophistication, wide applicability, and likelihood that they keep evolving. Some of these technologies, such as radar and catalytic chemistry, have been around for generations. Many have already found commercial application. Some, such as pharmaceuticals and packaging technologies, are broad terms but their horizon is still unlimited, so they're on here as well. Others are newly released from the lab. The newest ones, such as wavelength division multiplexing—a way of simultaneously sending signals at different wavelengths (that is, different colors) through the same fiber-optic cable—have yet to prove their commercial viability, but I believe they eventually will.

Shoreline's Picks

Our hypothetical friends at Shoreline review this list, and from it get some ideas to explore further. First they decide that *small solid-state lasers* are an ideal way to begin to insert optochemical measurement technology into their instruments. That choice also provides some screening criteria for their acquisition effort. They also realize that in light of their intensive efforts to dominate field testing of water, they should seriously consider incorporating *global positioning systems* into their handheld sensors so that their equipment can identify exactly where an observation is taken in the field.

Table 8.1: Cool Technologies With a Future

Materials Technologies
Ceramics
High-temperature ceramics
Food ingredient technologies
Polymer technologies
Plastics compounding
Fiber-optic materials
Conductive polymers
Nanodispersion
Electrochromics
Powdered metal alloying
Fiber reinforced composites
Designer molecules
Composite fibers and textiles
Synthetic fabrics
Lasers
Performance coatings
Adhesives
Packaging technologies
Labeling technologies
Battery technologies

Telecommunications Technologies
Code division multiple access
Wavelength division multiplexing
Wireless technologies
Low-earth-orbit satellites
Global positioning systems
Internet communication and commerce
Cable multiplexing
Video conferencing
Radar
Remote diagnostic technologies

Design Technologies
Ergonomics
Computer-assisted design
Computer-assisted manufacturing
Computer-assisted software design
Factory automation

Decision Technologies
Geographic information systems
Mathematical algorithms and analysis
Measurement and test routines

Decision Technologies (continued)
Digital imaging
Fuzzy logic
Neural network analysis
3D image generation and conversion
Video imagery analysis
Project and program management technologies
Ultrasonic imaging technologies

Biological Technologies
Biochemistry and medicine
Pharmaceuticals
Biotechnology
Genetic engineering
Cloning
Biomechanical engineering
Noninvasive medical testing
Minimally invasive surgery
Gene splicing

Information Technologies
Microprocessors
Silicon chip technologies
Flat-panel displays and signage
High-definition TV
Control systems
Digital printing
Ink-jet printing
Synthetic imagery
Bar code technologies
Machine vision systems
High-capacity data storage
Data security (cryptography and data encryption)
Data compression

Process Technologies
Catalytic chemistry
Robotics
Fluidics
Desktop manufacturing
Systems integration
Biocomputers
Materials-handling technologies
Purification technologies
Electrochemistry

The new Shoreline products will thus be able to guide the user back to a specific point at pond side or a brushy, hard-to-locate stream site or even in the open ocean.

Shoreline managers realize they had better develop a way to either store their field data in their units or communicate the data in real time via *low-earth-orbit satellites* to a central site. This need for *storing quantities of data* brings up the idea of selling data-archiving services to organizations and agencies that want to store massive amounts of water-observation data, as backup, off site. Furthermore, Shoreline conceives of small, inexpensive remote units that broadcast continuous readings of chemical concentrations and meteorology to a central data-processing and analysis site. This brings up the issue of remote, *robotic sampling* to take and record samples of water for later collection and analysis.

The more the Shoreline team gets going, the more new ideas they think up. Suddenly, the no-growth future they have been contemplating is gone like a summer shower, forgotten in the rush toward the plentiful new possibilities they now see ahead of them.

This look at Shoreline should demonstrate that several of these cool technologies have wide applicability and thus might be useful in your business, even if at first they may appear unrelated to it. Full explanations of each of these technologies are available from standard encyclopedias or *The McGraw-Hill Encyclopedia of Science and Technology*, or—even more fun to look at—David Macaulay's *The Way Things Work.*

A Closer Look at Some Cool Technologies

A number of these cool technologies warrant further comment. For example, *labeling* technologies and *adhesive* technologies find application in an extremely broad range of industries. Developments in adhesives, for example, have revolutionized the labeling of fruits and vegetables, which for years posed difficulties. Fruits and vegetables are hard to label because they have so many different kinds of surfaces. Tomatoes have thin, tough skin surrounding a fragile interior. Banana peels have a leathery finish. Watermelons and cucumbers are smooth, while avocados have a pebble-grain finish. Apples outgas ethylene through their skins. And so on. But today special adhesives hold the labels onto the produce during shipment and allow for easy removal without damaging the food. Could your business benefit from some kind of new label on what it sells?

Bar codes have, of course, revolutionized various aspects of retail trade and industrial materials management. They are also finding application on the highways, in billing mechanisms for toll-booths. A car or truck can pass through a toll booth without stopping while an electronic reader scans its bar code and debits the relevant account. Any situation that involves counting and tracking can probably benefit from bar codes. Does anyone in your company still take inventory by looking at things, counting them, and writing down the result? Bar codes might be the answer.

Nanodispersion has broad application even though still in infancy. The process involves making extremely small particles by shooting streams of granular particles at one another at supersonic speeds. The particles strike each other and break up on impact into ever smaller particles, nanoparticles, that are only billionths of an inch in diameter. These particles can be mixed into virtually any substance for use in coatings, adhesives, medicines, papers, metals, liquids, plastics, or cements. The resulting particles collectively have tremendous surface area for their volume. So with the addition of a tiny amount of active ingredient, the host materials can gain unusual special properties, such as solubility, that they could not have if a larger amount of the ingredient were added. Nanodispersed particles enable their host materials to be used in completely new and different ways, such as adding unusual properties to thin plastics without weakening them.

Serendipity sometimes helps. Long ago, some technologist in the grip of an unexpected brainstorm introduced lead into glass. The result was a different, more valuable material: heavier glass with the same basic uses as regular glass but with much greater refractive capability. A craftsman could cut this new glass to look like a precious stone, to capture and refract light. This kind of "Eureka!" moment is where your tour of technology should take you.

The entire technology of *purification* yields a wide range of benefits. In fiber optics, it produces very high-performance glass. The purity of the glass fibers largely determines the performance of fiber-optic cable. In general, purification, especially that achieved by electrochemistry or separation membranes, enables us to pursue extreme performance on the one hand and, on the other, rid materials of contaminants. The purity of feed stock materials is a major consideration in pharmaceuticals and in microprocessor manufacturing. Hyperpure medications minimize side effects and deliver precise results. In many sophisticated manufacturing processes, the purer a critical material, the higher the performance. Purification technology is moving toward purity measurable in parts per trillion, and for some semiconductor manufacturing applications parts per quadrillion (PPQ). One PPQ is the equivalent dilution of two grains of table salt in a railroad tank car of water.

Imaging and image analysis are examples of what's possible when you combine several technologies. If infrared, ultraviolet, or sound images are first captured through tiny "cameras" and patterns in their images are then analyzed using high-speed logic combined with pattern recognition and other analytic routines, visual analysis can make a tremendous leap upward in efficiency and accuracy. Pap smears are now examined for cancerous cells by human inspectors using microscopes, a method prone to ambiguity and error. Better to use optical scanning and computer image analytic routines that assess shapes, edges, contrast, tone, color, patterns, density, and so on to identify rogue cells. Larger-sized samples could be examined and assessed more quickly while minimizing human factors such as inattention, boredom, fatigue, and inexperience. In general, imaging and image analysis have wide application in health care, as well as in manufacturing, for positioning control, monitoring, and quality control; in security for identification and motion detection; and in remote inspection such as within confined spaces in machinery, underground, and space exploration.

Microprocessors, of course, are everywhere. When these chips are coupled with a technology called micromachining—extremely fine chemical or mechanical milling used to produce objects measuring hundredths of an inch across—an almost infinite array of sensors and switches can be made. One recent development is a baseball whose cover contains a radar-on-a-chip that clocks the speed of a pitch, enabling every sandlot hurler to try for ninety-five miles per hour. Aside from measuring speed, these kinds of micro devices can detect temperature (to trigger heating or air conditioning), impact (to deploy airbags), light (to adjust lighting), or virtually any other physical characteristic. Such chips can not only set processes in motion but also capture information. Large detectors now doing these jobs are being replaced by chips the size of a fingernail and smaller.

Having seen these technologies from a broad, top-level perspective, you may still be asking yourself, *How can my company make money with them?* There's a simple and direct way.

Think Like an Integrator

Most businesspeople approach technology from the perspective of the technologist, the developer of the technology. They often seem to feel that only the developer can legitimately use a technology, that licensing and using someone else's technology is akin to plagiarizing.

A better posture is to think of yourself as the universal user of technology. That is, think like an integrator. Think of how your products and services meet your customers' needs. Then think of how much better they *could* meet those needs if you were to use technology to enhance their effectiveness, boost their efficiency, or otherwise transform them for the better.

A master integrator of technology acts as the architect who chooses from all the good sources and suppliers to design a construction project for his clients—or as the symphony conductor who creates works of artistic beauty by training the best individual performers to work in a cohesive, smooth-functioning whole. The conductor does not avoid works with drums in the score because he's not a great percussionist. No, he engages the best percussionist and blends that individual's outstanding skills with those of the other musicians to achieve the best possible results. So, too, should you approach technology.

Let's take our tour of technology up to beautiful Idaho to see a simple example of technology integration in material substitution. We'll stop in at TJ Industries, formerly Trus Joist, which since its earliest days in the 1960s has been an inventive technological pioneer. Trus Joist had a line of wood and metal truss-shaped joists used to support a flat roof above and a ceiling below, as shown in Figure 8.6. In Trus Joist's proprietary design, cheap, plentiful, western wood two-by-fours replaced expensive eastern steel for the top and bottom pieces of the joist.

But Trus Joist's original structures were expensive to build because they required labor-intensive assembly to pin the metal zigzag members together with the wood to form the web, or verti-

Figure 8.6: Trus Joist's Original Design

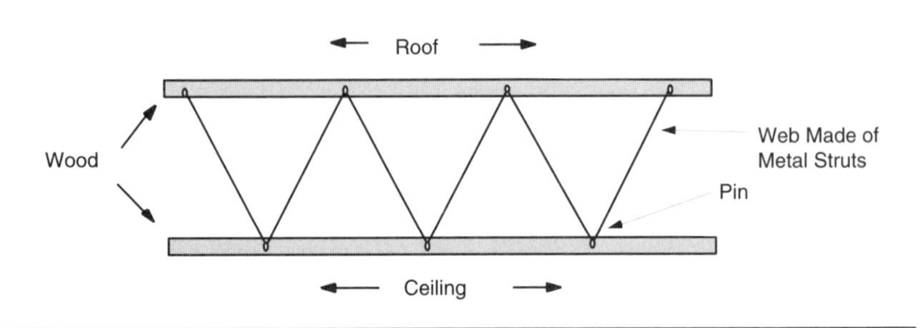

cal part, of the joist. To overcome these cost problems, the company invented its own solutions and borrowed a number of other companies' wood technologies in its redesigns of the trusses. These redesigns replaced the metal components with several products from private and public research labs. First, the metal struts were replaced by a one-piece web made of *wafer board*, which is flat paneling made of flaked waste wood. Later, the by-now-expensive wooden two-by-four horizontal beams were replaced by a new product called *stranded wood*, which is two-by-fours made up of long, very thin strips of waste wood glued together. By means of its technological awareness and its inventive genius, Trus Joist has gone through several generations of products, integrating the wood and adhesive technologies of others into a lightweight, space-saving, shallow, easy-to-make, and easy-to-use joist.

The new materials made with its own and borrowed technologies has enabled the company to eliminate metal from the product line. Not only are its costs lower, but the rigidity and strength of the composite woods equal those of metal and surpass those of more expensive noncomposite wood. Trus Joist combines its own new, improved raw materials with ideas from vendors, with whom it works closely. Through several mergers, Trus Joist has grown to be a $700-million-plus business named TJ International, with operations in the United States and Canada.

While we're in Idaho, let's look at another example of integrating technology into an operation; this example shows how technology reduces labor costs and transforms a business. Twin City Foods was a pea cannery thirty years ago. With the rising popularity of frozen foods, Twin City transformed itself into a large frozen pea packer step-by-step. For example, the company used to perform final quality control by hand. The former plant design routed the peas coming out of the shelling machinery onto a thirty-inch-wide conveyor belt. The peas then were conveyed about forty-five feet on the belt between two rows of inspectors, standing on both sides of the belt. These inspectors, about fifteen or twenty on each side, would visually inspect the peas as they passed by. They'd pull out the culls and the pods and the off-color peas, which Twin City then sold to local farmers as animal fodder. The rest of the peas would go off the end of the conveyor belt into the "good pea" trough and travel on for further processing and canning. That was Twin City's inspection capability in the 1960s and early 1970s.

As a technology integrator, Twin City revolutionized this labor-intensive inspection process by applying several technologies to it. The quality-control inspectors in hairnets have been replaced by lasers and software in neural nets. The peas travel a conveyor belt even today, but they move much faster now. Inches before the peas reach the end of the belt, they are scanned by a laser and the image is interpreted by neural net technology. The neural net incorporates an analytic routine that creates and categorizes combinations of characteristics, as a human inspector would. For example, if image analysis and color spectroscopy indicate that the scanned object is green and shiny and regularly shaped and smooth and the right size and so on, it passes—it's a good pea. If it's all of those things, but not shiny and not smooth, it fails—it's a small pod. If it's all those things, but too small, it fails—it's an immature pea. Thus the system can identify the good peas and differentiate them from everything else. All the rest fail the test.

A few inches further down the conveyor belt, past the laser scanner and interpreter, all of the peas—the good, the bad, and the ugly—go off the end of the belt and fly in a gentle arc toward the good-pea trough. But partway through their gentle flight the pods and the uglies and the off-sized peas and the debris are literally shot down from above by tiny, momentary jets of compressed air. The culls land in the fodder bin lying below. This ingenious high-speed sorting is possible because of integrated technologies: the laser that scans the peas; the neural net software that interprets the images on the belt; a computer that instantaneously acquires, stores, retrieves, interprets, and communicates the data; and the overhead bank of tiny switch-operated compressed air jets at the end of the belt that blow the problems away.

The new Twin City Foods can process far more peas with virtually no errors at a faster rate with minimal labor costs. Gone are the forty-odd inspectors, who probably tired of their job quickly anyway. In fact, this application of technology along with many others in freezing, packaging, measuring, handling, and labeling, enabled Twin City to transform itself into a high-volume frozen pea packer, selling frozen peas in packages weighing up to three-quarters of a ton, giant freezer packs for repackaging and labeling by other companies. The footprint of the plant is substantially unchanged, and of course the raw product, peas from the field, hasn't changed; yet the economics, scale, and focus of the business have changed completely. Twin City is now a wholesale packager of peas and a private-label freezer. Using technology to transform its operations and output, Twin City has concentrated on several specialized functions.

Twin City's role as a technology integrator is another case of applying sophisticated technology—notably a tremendous amount of complex computing that occurs in the interval between the moment the peas are scanned and the moment they are either shot down or allowed to cross the gap to the good-pea trough—and applying it to what most of us think of as a pretty basic, low-tech business.

Back on the Highway

Let's leave Idaho and get our tour back on the open road. The major players in the trucking business, including carriers, fleet owners, and truck and aftermarket equipment manufacturers, are all transforming their vehicles with technology. Trucks are becoming platforms for a surprisingly wide array of electronic technologies. Cellular communications to keep drivers in touch are already available. Soon long-haul trucks will have geographic information systems capable of communicating their exact location to both the driver in the cab and the home office of the carrier that operates the truck. Before long, perhaps by the time you read this, electronic sensors will monitor the position and movements of the driver as he's driving and set off a chirp alarm if he stops moving, signaling that he's falling asleep. Some monitor systems already sense, record, and transmit the temperatures of refrigerated loads to let shippers know how "delicately" their cargo is being treated.

From the truck driver's viewpoint, a tractor-trailer is a collection of blind spots hurtling down the road. For example, the area immediately next to the right-front fender of the cab is very hard for the driver to see. So collision avoidance radar will be positioned there and at other points on the truck where mirrors and viewing windows can't do the job. Radar with an effective range of up to eight feet can set off a chirp alarm to warn the truck driver of that little Geo Metro to the right in the next lane. What does this mean for trucking? Fewer accidents, fewer damages, fewer lives lost, lower insurance rates, faster deliveries, and happier customers.

Incidentally, Cadillac has already developed a system that automatically places a cellular call if the airbag is deployed, to a central station staffed twenty-four hours a day. The cellular phone interacts with a global positioning system and the airbag sensor to tell the monitoring office to send help immediately. Trucks, too, will soon have such systems.

We're *All* in Technology Now

If you think of yourself as a technology integrator, you'll recognize all that you can do to create products and processes to meet your customers' needs and solve their problems. You happily draw from a broad range of technologies and put them into service in your business. The technology that can help you need not be at the core of your business. Making joists, freezing peas, or scheduling trucks might be at your core. But incorporate technologies like the ones in these examples—new materials, new quality-control methods, new productivity tools, and improved safety devices—and you create value for your customers that produces TechnoLeverage for your business, without fundamentally changing the business you are in.

Clearly, you need not be a technologist—or even have any on staff—to be able to do this. As an integrator, you need to be aware of the technologies that are out there, as we've seen here and in Chapter 3. Next, devote effort to thinking of how technology might work in your products, processes, or markets. Then integrate them into your company and its operations.

The meaning of the statement "we're *all* in technology now" should be clear. This is where our tour of technology has brought us. We've driven away from thinking of your business as a low-tech, low-growth, or single-technology operation. We've looked at several old-line processing businesses that have changed by using technology. Adopting other people's technology can significantly accelerate your company's growth, returns, and competitiveness.

To achieve TechnoLeverage, companies in so-called traditional or low-tech businesses must constantly seek ways to apply technology to what they do. Meanwhile, companies actually in a technology business must constantly renew their core technologies. Given that technology now drives business, is there really any other path to pursue?

Review and Preview

This chapter has shown how to assess your internal capabilities, the technology and skills you need to reach your business goals. Our example of Shoreline Sensors shows step-by-step how to assess your current situation.

We've looked at a number of the technologies that are out there in the economy and at several companies that have pressed them into service, often in surprising ways. This chapter's tour of technology has put you on the road to understanding how technology—broadly defined and shrewdly employed—can boost any business. We've noted TechnoLeverage Tip Number Eight: that *awareness of many technologies greatly improves your choices of new things to do.* Awareness of technology is fundamental to success at TechnoLeverage.

Technology generates change. In Chapter 9 we look at the kinds of change technology generates for companies, and at the best ways of managing these changes from an organizational perspective.

CHAPTER 9

Future Perfect

Creating an Organization That Creates Change

All of us in business have heard a lot about technological change: its speed, inevitability, and constancy. Technological change deeply affects every business organization. Yet managers of businesses that create new technology, apply it, sell it, and use it receive little practical guidance on dealing with organizational change caused by technology. Even *Dilbert* suggests that technology companies could manage their organizations better. This chapter fills the gap.

Managers must answer two basic questions about technological change: "In which direction should we go?" and "How do I get my people moving in that direction?" Knowing the right direction is critical because technology offers no time to reverse course and catch up. You want to head toward the right place and be there at the right time. Getting your people moving—together—is essential because their collective and individual efforts determine whether or not your organization becomes what it must be to survive and prosper with technological change. Moreover, you cannot order people to change. You must lead them in the right direction and help them follow you.

Many of us think of managing technological change as coping with external sources of change—in costs, competition, computers, communications, regulation, and the economy. To be sure, the environment generates ample technological change. However, as a company uses technology to improve performance, it must deal with the internal changes that the application of technology generates.

A company using technology strategy automatically creates change. If you use technology to solve a customer's problem, differentiate your product, or gain competitive advantage, you ignite changes around and within your company. These changes in turn set off other changes that you must manage.

My approach to managing organizational change, particularly changes that flow from technology strategy, is straightforward:

- Anticipate potential changes.
- Choose the best course.
- Implement the changes.

That's the general outline. Although it sounds simple, we all know it isn't, so this chapter covers each of these steps in depth. Briefly, I recommend that you *anticipate potential change* with two simple tools: the TAS and the technology adoption-diffusion model, first introduced in Figures 1.1 and 3.3. Because the obstacles to a new application are fairly predictable, you can use these road maps to determine where you are and where you're headed. The main focus of this chapter is on what you can control: your operation and how it is affected by your strategy for using technology.

The basis on which to *choose the best course* should be your resources and the coming demands upon them. Often managers misjudge both these elements and thus whether their company can accommodate the change it initiates. As a result, those managers' efforts are doomed from the start. By using the TAS and the adoption-diffusion model, you as a manager know clearly what you are getting yourself, your employees, and your shareholders into.

The issue of *implementing change* can be broken into manageable chunks. In the world of technology strategy, there are primarily two types of organizational change worth making: those that create new businesses and those that produce technological innovation. We therefore approach implementing change in terms of these two goals and discuss approaches and structures that help organizations accomplish them.

Look Ahead and Anticipate Change

The company pursuing TechnoLeverage can anticipate certain changes because of the nature of a successful technology application. Recall that at different periods in its lifetime an application is characterized as a unique, an exotic, a specialty, or a commodity. Most successful applications go through each stage as they achieve broader market acceptance, as depicted on the TAS.

Each stage of an application's life makes specific demands on the company that provides it. An exotic application calls for a different organization than does a commodity application. Thus for a given application, the TAS provides a map revealing both where a company is and where it is headed. Figure 9.1 shows the characteristics and challenges that applications at various stages of development present to an organization.

Using the TAS to See Change Coming

To assess the changes ahead, you must first know which category best describes your application and how your company's resources and approaches stand in relation to the application.

For instance, Table 9.1 shows that the labor content of most exotics is relatively high. If you produce an exotic, your corporate mission is to gain acceptance for the product. This in turn requires managers who can proselytize for the product. They have to be comfortable spending their lives surrounded by people who are skeptical of what they sell. Your organization needs flexible administrative practices and innovative sales approaches in this customized environment. Your developmental engineers must design products that fit the market's needs better than your initial model did.

Begin by examining your applications and your organization in light of what the table shows. Assess your current position and then look ahead. As you do, consider three questions:

Table 9.1: Organizational Challenges of Various Applications

Product Characteristics and Departmental Approaches	Application Type: Unique	Exotic	Specialty	Commodity
Example application	Electrically darkening window glass	Cellular phone electronics	Specialty polymers	Photocopy paper
Relative labor content	Very high	High	Medium; declining	Very low
Corporate mission	Produce usable products	Gain acceptance	Broaden awareness of utility	Maintain high sales volume and low costs
Best style for senior management	Explorers	Proselytizers	Penetration builders	Centralizers
Best style for administration	Project management	Flexible; experimental	Interpreting data; budgeting	Cost cutting
Best style for sales	One account per product; intensive hand-holding	Key account; innovative approaches	Finding new uses and customers; finding vertical markets; leaving no stone unturned	Inside order taking; great service for large customers
Best style for manufacturing	100% custom; made to order	Customized	Proliferating new products from basic designs	100% standardized production
Best style for R&D/engineering	Nobel Prize aspirants	Developmental engineers	Application engineers	Industrial and production engineers

1 Where are your applications now?

2 Are you supporting your current applications properly?

3 What changes lie ahead?

First, *where are your applications now?* If you have more than one application, do you as a company straddle two or more of the four categories? To benefit from Table 9.1, you *must* understand whether you are producing and selling a unique, an exotic, a specialty, or a commodity. If you have not already done so, figure this out immediately.

Second, *are you supporting your current applications properly?* If you have an application in a specific category, the table shows which organizational attributes you need to support it. That is, you must understand the corporate mission and have the right senior management, administrative support, and sales orientations. If you have organizational problems now, the cause may well be misalignment between your application's position and your mix of resources and approaches. As you might imagine, achieving alignment is harder if you have applications in more than one category; this is an issue that we address later.

Third, *what changes lie ahead?* If the application is successful, if its production and sales volume keep growing, it eventually becomes more commonplace and so moves rightward to the next zone. This means you can look ahead to the next zone on the TAS (unless you're selling a commodity) to see the demands and changes you face as the application matures. You can anticipate change by looking for the factors that accelerate movement rightward: more competitors, backward-integrating customers, more effective substitutes, and intense price pressure. Absence of these factors tends to stabilize applications and retard their rightward movement toward commodity status.

For example, if you have a successful exotic that is moving into the specialty stage, you're heading into new territory. You have to produce more units, and there are more competitors, so your labor content and customization must decline. The corporate mission shifts from raising marketplace awareness to increasing penetration. Sales must find many new uses and new customers for the application and penetrate vertical markets as well. This gives rise to product clusters in manufacturing and the need for application engineers.

As challenging as these changes are, you can anticipate them and sequence their impact on your organization. You can evaluate your people and help them prepare for the changes ahead.

A Closer Look at Sales Challenges

Given the prominent role that sales and marketing play in developing new business, let's examine the sales challenges at various stages of market development. For this assessment we examine the market, rather than the application, by using the buyer segments from the technology adoption-diffusion model in Figure 3.3: innovators, early adopters, early majority, late majority, and laggards. (These five segments broadly relate to the four application zones. In

Table 9.2: Sales Challenges in Various Market Segments

Product Characteristic and Departmental Approach	Market and New Buyers				
	Innovators	Early Adopters	Early Majority	Late Majority	Laggards
Market condition	Introduction	Acceptance	Growth	Maturity	Decline
This segment's percentage of market	2.5%	13.5%	34%	34%	16%
Key sales task	Find first customers	Broaden sales base	Establish complete account coverage	Maximize market share	Maximize sales dollar productivity
Size or type of sales force required	1-3 people	5+ people	Many people	Multiple sales teams and structures	Shrinking sales force
Type of salesperson	Highly entrepreneurial individualistic; a "first account" ace	Good in small teams; a gifted teacher	Organizer and account penetrator	Persuader and comfort provider	Hard-nosed; nuts & bolts orientation
Sales message	"This is brand new and offers high performance."	"New, but proven."	"Everybody's doing it."	"You'd better catch up."	"You have no other choice."

general, innovators begin purchasing at the unique stage, early adopters begin purchasing at the exotic stage, the early majority buys at the specialty stage, and the late majority and laggards begin buying during later-stage specialties and commodities.) Table 9.2 shows the sales challenges organizations face at various stages of market development.

Clearly, you cannot sell an exotic as you would a commodity. Everything, from the sales task to the type of salesperson to the sales message, differs radically. The best plan builds on recognizing that each stage of market development requires a different sales and marketing plan. You need different plans, each aimed at converting groups of customers with varying objections, concerns, and perceived risks.

As we learned in Chapter 3, the technology adoption-diffusion model was developed by the U.S. Department of Agriculture in the 1930s. It was expounded further in several editions of *Diffusion of Innovations* by Everett M. Rogers, first published in the early 1960s. That book defines the first 2.5 percent of the buyers of a new product as innovators, the next 13.5 percent as early adopters, and so on, as the adoption-diffusion model figure shows. It's a proven process by which individuals and organizations explore, accept, and buy new technologies.

Using technology strategy means dealing with these issues as they arise at each stage of the adoption sequence. Failure to address them can undermine your efforts to grow and hand a more savvy competitor an easy opportunity.

Like Table 9.1, Table 9.2 helps you assess the changes ahead. You can see how your sales force has to change key tasks as well as its size, approach, and overall message. Having undertaken this analysis, you are now prepared for the next step: deciding what organizational changes to make.

The Way to Go: Choosing the Best Course

Managers of a rapidly growing business often find themselves with the wrong resources for the next stage of their market. One day they realize, "Nothing we do seems to be working. Where did we go wrong?" Often what "went wrong" was failure to consider how the organization should be structured in light of the forces affecting it.

Two Roads Lie Before You

Unless you sell a product that by some miracle remains in one zone of the TAS for an extended time—as color film stayed a specialty for fifty years—you've got to keep generating new applications and winning new markets for your technology.

Few products, however, simply stay in one category. Like the markets for microcomputers, sugar substitutes, cell phones, floppy disks, fiber-optic cable, bar codes, cable TV, and many other applications of technology, the market for your products probably changes constantly. Market forces eventually pull all successful applications rightward on the TAS. As they do, you face declining margins and the challenges associated with moving rightward in Tables 9.1 and 9.2.

When you see your margins falling, customers gaining power, and competitors marching in, when you hear that someone is adding significant capacity or you notice that customers perceive diminishing value in your products, should you go with the flow, allowing your organization to move to the next application stage and market segment? Or should you resist the current and move back up the TAS by developing a new, higher-value, higher-margin application from your technology?

Since moving either rightward *or* leftward on the TAS entails significant organizational change, you must honestly assess the resources you have and those you need. You must honestly assess your people's skills, and you must carefully anticipate your product's prospects in the new direction, about which you may know little.

For instance, if your unique or exotic applications show real potential for sales beyond the innovators or early adopters, you must stop and think. To move beyond these customers, you must change dramatically. New people—new kinds of people—are necessary in production and in sales and in marketing. You have to expand production: Good-bye custom manufacturing, hello assembly line; or perhaps good-bye Made in USA, hello Made in China. You have to change your sales and marketing approaches: Good-bye entrepreneurial enthusiasts and technology teachers, hello professional sales managers.

Can your company make changes of this sort? Does it really want to? Should it try? You must know the answers before you choose to change.

The Toughest Move of All

It is difficult for most companies to straddle two application categories, so I do not recommend trying. Actually it's next to impossible for a single organization to straddle more than two. There are creative ways to structure a company for this kind of change, but not as a true *single* entity. We look at these structures later in this chapter.

One way or another, many companies do manage the changes involved in moving from unique to exotic or from exotic to specialty. Some even move smoothly from unique to exotic to specialty. But relatively few businesses move successfully, let alone smoothly, from specialty to commodity. This is the toughest change of all.

Consider this: Once a growing application is well into the specialty stage, it has sold well to early majority buyers and may be nibbling at the late majority. However, at this stage it faces new, more intense pressures. Price pressure is eroding margins. Competitors have launched parity products. (Commodities *are* parity products.) Worst of all, your commodity-zone competitors have experience in producing and selling in high volume at low cost, and they may have strong name recognition to leverage as well.

Many managers find themselves seriously tempted to expand their operation to handle volume production of commodity products and fight for share. The mass market beckons. The company would get much larger—and most corporate and personal reward systems aim for larger size. These markets seem gigantic, open, endless. They sing a siren's song, encouraging companies to "make money on volume, not on value."

To play the volume game, your organization has to become a commodity producer. This means walking away from most of what made you successful when you were making money by adding value: the highly educated people, the smaller scale, the targeted approaches, and the sophistication. All the costs you've borne to create and make higher value-added products are now excess baggage weighing you down. Inventive technologists wither and die from lack of challenge in a commodity environment. Also, the asset-intensity of a high-volume, low-margin operation makes such operations difficult to finance internally. Commodity production needs capital, probably from a public offering or sale of additional stock, and perhaps debt. Ultimately, you have to change almost everything about your company to conform to the commodity column of Table 9.1 and the late majority and laggards columns in Table 9.2.

A company considering this move has the best chance of prevailing if it has barriers for protection. They include regulation, patents with useful remaining lives, and access to markets strengthened by strong contractual or personal relationships. The higher the barriers, the better. A strong brand helps, as does an established—one hesitates to say "loyal"—customer base. A brand, together with a barrier such as access to markets, can do wonders, as Microsoft has proved. However, as Apple found out, a brand hardly guarantees success. Attractive economies of scale and ample resource availability also help.

If cold calculation tells you that your company can change into a commodity producer, fine. Yet far too many managers stumble into it, blind to their company's weaknesses, unduly confident of its strengths, and dimly aware of the underlying market dynamics. Seeing their product becoming commoditized, they invest more money to chase *inherently* diminishing returns. Then, when "less sophisticated" competitors who have the low-cost, high-volume game down pat overwhelm them, they're shocked. Above all else, the commodity zone rewards comparable products produced at lower costs.

Going Back (Not Forth) to Growth

What if your assessment tells you to stop right there, wherever you are? What if you decide to leave commodities to commodity producers? What if you sell an exotic and you choose not even to enter the specialty zone? Are you supposed to hang up your gloves? No! You are supposed to play to your strengths: your innovative people, your advanced development-and-production capabilities, the technological excellence of your products, and your problem-solving abilities.

Playing to your strengths means taking your organization leftward across the technology applications spectrum. How? By pursuing new business based on applying innovations. This takes the company in a circular or elliptical path through two or perhaps three application categories. For example, if you are based in the specialty zone and want to move your company back up the curve, first look for a unique application that may be ready for acceptance by a larger market. Then make it more widely acceptable and introduce it to more customers, making it an exotic. If you already work in the exotic zone, you can devise something completely new and then introduce it at the unique stage. With the proper organizational structure, you can repeatedly cycle from unique to exotic to specialty and back up to unique. This can be a very profitable alternative to pursuing commodity production.

However, to pursue new markets and innovations that take you back up to higher-value, higher-margin applications, you need to make changes in your organization.

Now Do It: Implement Change

Let's look at some nuts-and-bolts issues in managing two types of change: launching major new business efforts, and achieving technological innovation. Then let's examine the organizational structures that promote these changes.

Here's the key implementation point, TechnoLeverage Tip Number Nine:

> ▶ *TIP NUMBER NINE:* **Small work groups are the main tool for achieving technological change in business.**

Everything about small work groups fits the technology environment: their flexibility, camaraderie, speed, responsiveness, openness, and their experimental urges and inherent impatience—even their mild sense of loneliness and isolation. They're a natural place for the can-do entrepreneurial inventor-scientist. Everything we know about achieving technological change underscores the value of small groups.

Business entrepreneurship and technological innovation usually spring from individual efforts. In a business organization you can—you must—channel those efforts toward a collective goal. You must balance people's individualistic, inventive, and entrepreneurial needs against their social, cooperative, and collaborative needs. This balance can be struck most easily in small work groups. Small groups provide structure and affiliation and still allow for individual effort, accommodation, and recognition. Small work groups see their goals more clearly than large organizations do. Small groups spend less effort on politics and "group process" and direct more time and energy toward the group's goal. Shaping and directing small groups are your critical tasks for implementing technological change.

Many executives have commented on the difficulty of getting an unwieldy "corporate battleship" to change course. The unwieldiness stems from the conflicts inherent in a large group. Small groups encounter less friction because they have less mass, because their members' roles and responsibilities are clearer, and because each member's contributions are more visible. Those who can't contribute—or who can't be trusted—quickly drop out or get forced out. This winnowing helps people in the group see and respond to opportunities and threats faster. The simple ability to take action makes small groups effective catalysts for change. If you have to make your corporate battleship nimble, get out your cutting torch and begin carving it into several lean and maneuverable destroyer escorts.

Keep these thoughts in mind about small groups as vehicles for change as we look at new businesses and innovation.

Structuring a New Business for Change

Building a new business is a manager's toughest task. I refer to building a completely new business, not simply finding new customers. New businesses are risky, unknown, untried, unproven. Their leaders swim in a sea of uncertainty. Most of the people involved have to learn new skills. No one knows the new business as well as he knows the existing one. New businesses need leaders who can act capably in the face of uncertainty.

The new business disturbs relationships among various company functions. Research and development, finance, production, marketing, and sales have their own agendas regarding the new business effort. Meanwhile, you want change, not chaos.

The ideal individual to develop a new business possesses an uncommon mix of characteristics. The leader should be an entrepreneur and risk taker who also knows how to be part of the corporation. This kind of person is essential to the success of most new business efforts.

Here are two examples of how to develop a new business organization, one bad, one good, both from the paper industry.

Our *bad* example is a $300 million manufacturer of writing and printing papers, a division of a multibillion-dollar forest products company. This manufacturer's goal was to start selling directly into retail stores in addition to its usual distribution through commercial printers. Senior management gave responsibility for this effort to a marketing manager from the existing business—without removing his former responsibilities. He was designated a "coordinator" of the effort (not "leader" or "boss") and given the job of coordinating a fifteen-person team, consisting of representatives from each function in the company. The man chosen had a peacemaker's style and tried to manage by consensus. He avoided rapping knuckles or speaking bluntly, even when necessary.

Each team member essentially represented his or her own function. Naturally the group became a "court" where any member could "retry" other decisions on unrelated issues that had arisen elsewhere in the company. Because of the leadership style and makeup of the group, each representative had veto power over any decision. Only unanimous votes could decide anything.

Senior management never communicated a cogent vision or set of expectations, nor a specific goal or timetable. Instead, top management just wanted "to get into those retail channels," and it began approving budgets for the effort.

After two years and $2 million spent on meetings, research, planning, design, packaging, inventory, and travel, there was still no top management commitment and not one consequential or repeat account. Total sales: a few thousand dollars

For our *good* example, let's look more closely at Abitibi-Price, the manufacturer of groundwood paper mentioned in Chapter 6. Senior management placed a vice president of market development and technology in charge of the task and very publicly supported his efforts. He was hired for this work only, and he and his one-person sales "team" were dedicated solely to this project. They were given a mandate to "develop a new line of business," and their progress was reviewed regularly. Manufacturing was told to give them what they asked for. They could call on any prospective account, anywhere.

Once the VP landed a customer for a new application for groundwood paper, the account was handed off to the salesperson, leaving the VP to pursue new business. He was the technological and performance "opener." He chose to build the new business one customer at a time, and he was able to do this thanks to his sales and marketing skills combined with his background in paper chemistry. He pursued business with engineers and product designers who could consider the new applications. He avoided calling on purchasing managers.

The result was $50 million annually in sales from the new business after four years—a highly profitable business in the world of groundwood paper, which is truly a commodity. Abitibi-Price achieved these sales in exchange for two reasonable salaries and plenty of airline tickets.

Do's and Don'ts of Using New Business for Change

A number of lessons emerge from these and similar cases: Strike a balance between starving the effort and smothering it in resources. Make goals clear without having the new-business people overcontrolled. Put the right person in charge of a small team dedicated solely to one effort.

Many senior managers undermine an initiative by asking people to develop an entirely new line of business as a "special project" in their "spare time." That's understandable. Management hates removing good people from their regular jobs, so this spreading of effort looks like a smart economizing move. However, this is false economy because it hobbles the new business effort. If you believe that new business is the company's future and that you're giving the job to the right people, then you must give them every chance of success. This means making new business a full-time job for the most knowledgeable, energetic, reliable people you have, and grooming such up-and-comers in the existing business.

Following are ten other guidelines:

Five Do's

1. Do form a special unit to develop the new business.

2. Do place one broadly capable senior-level person in charge of developing new business.

3. Do make sure that this leader has both an entrepreneurial sales background *and* proven technical abilities.

4. Do organize the effort so that its salespeople get maximum exposure to new customers and its technical people can respond directly and immediately to customers' inquiries.

5. Do build a culture that understands the urgency of the new business, and show genuine commitment to it at the top.

Five Don'ts

1. Don't rely on your existing sales force to sell truly new products (the most common mistake).

2. Don't permit more people than absolutely necessary to participate in new business decisions.

3. Don't hesitate once you identify an attractive market; opportunities are perishable.

4. Don't overfund or underfund new business development or load it up with costly people or assets that doom it to failure.

5. Don't condemn people for new business failures and thereby miss the valuable lessons to be learned.

Generally, err on the side of more freedom and less money. Assigning blame for failures is grossly counterproductive. If some effort fails, diagnose the failure thoroughly, make the appropriate changes, and then move on and try something else promptly. This is a trial-and-error activity, and the key word is *trial*.

You must allocate resources to activities that develop something to sell and then find someone to sell to. Money spent developing an application or a market is usually well spent, provided you watch for warning signs that a return might not materialize. If you see clear signs that it won't, cut your losses and move on to a more promising application or market. Spend frugally on research, studies, meetings, and imposing furnishings. Spend your cash on whatever directly helps you find customers, build the business, or develop a product. The best uses of new business funds are *product development* (including engineering, design, prototypes, concept tests, and product tests) and *market development* (customer prospecting, market tests, sales calls, and postsale support). Avoid anything fancy.

Managing Innovators

Truly innovative people cannot readily be directed in the command-and-control sense of the term. For one thing, they often have more raw intelligence than their managers, and they know it. For another, innovators function in their own reality, which they have constructed of intuition and diverse experience. The best managers of innovators set forth a clear goal—always as a challenge—and provide motivation, context, independence, and organizational protection for the innovators' activities.

A seasoned business manager realizes that true technological innovators, a tiny portion of a company's personnel, often lack the dollar-and-cents orientation of an experienced businessperson. Most innovators respond well to challenging goals, a stimulating environment, and plentiful support, respect, and encouragement. These things good managers can provide in abundance.

Keep Them Simmering

People crave both stability and stimulation. As individuals, they favor stimulation. Without stimulation, individuals become bored, unfocused, and resentful. In groups, however, they favor stability. Without group stability, people become fearful, disorganized, and unproductive. Therefore managers of new business efforts must continually strive for the right mix of individual stimulation and organizational stability.

This balance is best achieved by setting the right challenges before the right people. Think of people in a company as a saucepan of water simmering on a gas stove. Some people—the innovators and the entrepreneurs, a small portion of the staff—are steaming all the time. They hover above the outfit on a level of their own; they need lots of stimulation. The rest of the staff needs stimulation too. They are bubbling away on the day-to-day operational, marketing, and financial matters. However, they also crave stability and resist change. Their resistance to change gives the organization heft.

The flame under all this represents the organizational environment, which comprises culture, structure, management style, and incentives—all factors that affect people's performance. To get the right result, management must provide an environment that keeps the day-to-day people bubbling and allows the innovative, entrepreneurial individuals to expand, to become steam. This takes management flexibility. If the environment overheats, the outfit becomes unstable. The innovators lack structure and fly into space while the day-to-day people swirl around in gossipy confusion. But the alternative, placing a tight lid on the outfit, would stifle the innovators and leave everyone bottled up.

Broadly, you can create a challenging new business environment by constantly casting people's work as a mission, not a job. This calls for presenting tasks and responsibilities in terms of goals and results. These must be difficult enough to challenge people, but not so difficult as to frustrate them unduly. Casting work as a mission also calls for giving people flexibility as to the ways and means by which they reach the goals and get the results. This leaves them free to bubble and expand, but in a directed manner.

Flexibility should extend to incentives. As a company grows, it changes, taking on new kinds of people and losing some familiar faces. A company mixing scientists and technologists with marketers and production people needs an array of incentives. For instance, exploration and discovery motivate scientists and technologists. They enjoy confronting ever-greater technological hurdles, an activity frustrating to everyday businesspeople. Also, some technologists work better on structured problems than on unstructured ones. Smoothness of fit between work and worker and opportunities to use and build skills are major incentives to most professionals today, regardless of their specialty. It costs little for managers to supply such incentives.

In fact, many low-cost or cost-free incentives can be usefully targeted to various groups. Businesspeople typically enjoy recognition of their achievements in the eyes of the organization. Salespeople in particular enjoy public rewards, which many technologists would find unimpressive or embarrassing. Many companies could be more flexible in the area of "lifestyle balance," another low-cost way of retaining good people, particularly those who have families or relish travel. Access to the president and senior management is yet another nearly cost-free incentive that most companies underuse.

In other words, the tactic of segmentation works for your employees just as it does for your customers: Incentives can be offered broadly but aimed at specific employees. Remember that incentives aside from money can be powerful. People do not innovate or change solely in response to money. Inspiring goals, a challenging mission, a rewarding environment, and thoughtful, attentive management also encourage innovation.

Two Ends of a Corporate Continuum

The characteristics listed below represent two sharply contrasting organizational environments. Each element in each pair lies at the opposite end of a continuum. These extremes underscore the importance of creating the best possible environment, which is the key task in managing technological change:

Best	***Worst***
Rational	Political
Entrepreneurial	Bureaucratic
Analytical	Instinctual
Consensual	Autocratic
Flat-structured	Hierarchically structured
Forward-thinking	Status-quo–oriented
Focused on team values	Focused on self-preservation
Generous	Conniving
Encouraging	Critical
Inclusive, open	Exclusive, closed

The "best" characteristics create a flexible environment, which is more amenable to change and attracts the most entrepreneurial and innovative people.

The Shape of Things to Come

Our research into technology companies shows that their net profit margin increases with size until they reach $200 million to $250 million in annual revenue. At that point, net profit margins begin to decline. Smart managers break their organization into smaller units, "hiving off" small work groups as the whole organization grows. When using this tactic, management must find productive, logical ways to separate those small groups from the bureaucracy of the larger organization. Here are three broad guidelines:

1. Strive for singularity of purpose. Give the new work group a very specific, high-value opportunity to pursue.

2. Make the work pertinent, and structure the organization to the work required. One group might focus on a market, another could focus on a product, while the third could be a joint venture. Good technological organizations often use idiosyncratic structures.

3. Provide autonomy. Managers of business development must fight the managerial tendency to centralize control. Offer the new groups local control.

What follows is a discussion of several organizational approaches to achieving technological growth using small groups.

Separate Marketing Organizations

If the core technology of a company supports various applications produced by one central manufacturing technology, a hub-and-spoke structure (Figure 9.1) enables it to approach various markets effectively. This calls for organizing the sales and marketing (and perhaps distribution) functions along market segment lines. For example, a paper manufacturer may have one marketing and sales team for book publishing paper, one for magazine and journal pub-

Figure 9.1: Hub-and-Spoke Expansion Structure

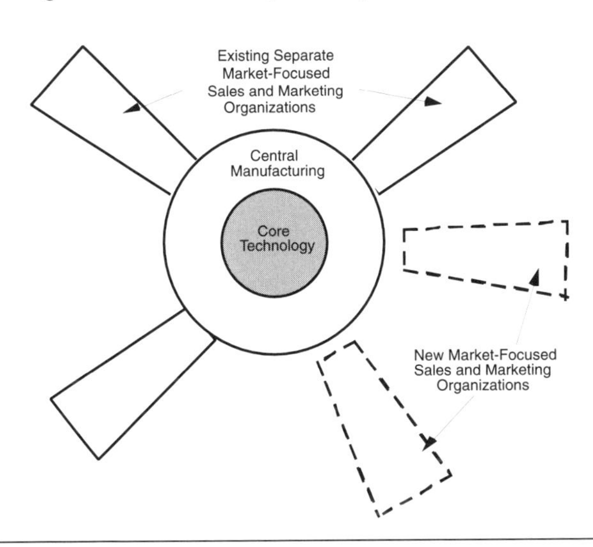

lishing paper, one for advertising and promotional printing papers, and one for financial services papers. If the company decides to approach a new market supported by the same technology and plant, it can add a new spoke and do so by starting small.

Separate Manufacturing Organizations

If a company's core technology can provide the ideas to support two or more separate manufacturing facilities for separate product lines, breaking up the expansion parts of the organization by branching can work well (Figure 9.2). For instance, a technology for coatings may enable a company to make both powder-based and solvent-based coatings, and it may be most efficient to have them produced in two separate plants. Each plant organization can then either support its own sales and marketing organization (markets one through five) or feed into a common sales and marketing function (market six).

A Team That Mimics the Technological Skills of the Parent (Replication Team)

A company may have various technical specialists contributing to a product. For example, a construction equipment manufacturer may have an engine team, a chassis team, a hydraulics team, a sheet

Figure 9.2: Branching

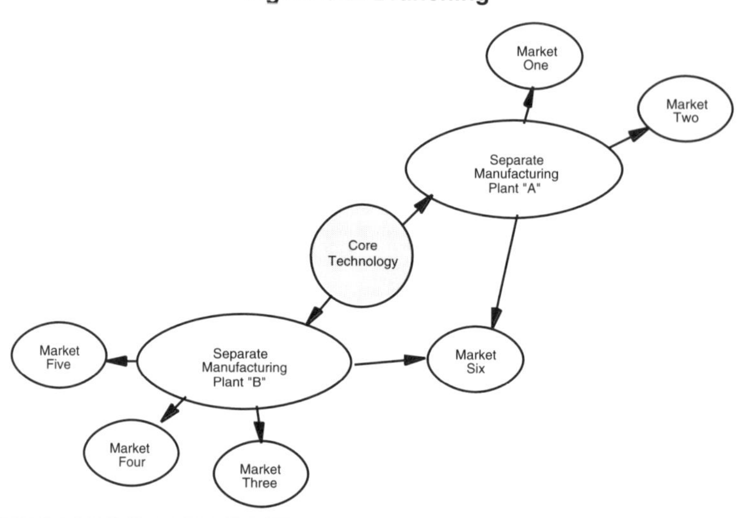

metal and cab team, and so on. These technical teams have separate specialties, which reflect the composition of the main business.

If such a company wants to start up a new business, it cannot make that effort the responsibility of just the engine people or just the chassis people. The effort requires a new team (Figure 9.3),

Figure 9.3: Replication Teams With Technological Skills

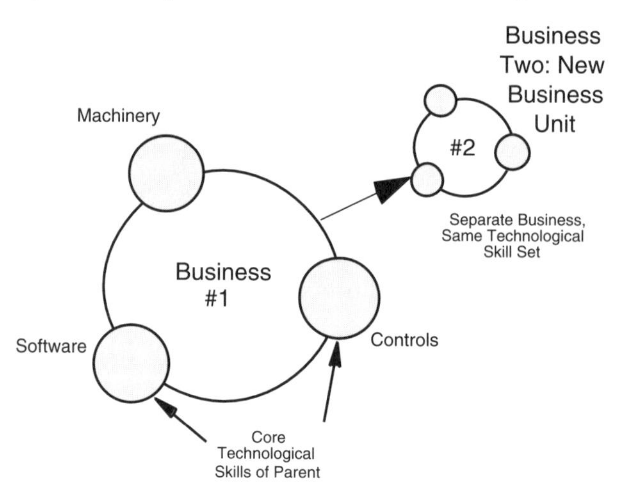

and the best candidates are a small group that replicates the technological capabilities of the company and applies them to excavators, elevators, or whatever the new endeavor may be. I like to think of replication teams as "garage shops." They more or less reproduce in microcosm the essential value-added of the larger organization.

A Team That Mimics the Functional Skills of the Parent (Application Team)

Application teams are similar to replication teams in that they parallel the essentials of the larger organization in a smaller one. However, an application team is drawn from the familiar internal functional disciplines rather than from technological areas. Most technology-based businesses are functionally organized (consisting usually of manufacturing or operations, sales and marketing, engineering or design, and finance or administration). So an application team might include an engineer, a marketer, an operations manager, perhaps a financial specialist, all in one workgroup, all focused on a new application or market or product (Figure 9.4). These functionally organized application teams are very commonly used. I think of application teams as a slice of the company, carved out to create technological growth.

New Groups That Regularly Use Developments of the Main Technology (Satellite Groups)

When a company has an extremely rich core technology giving rise to a stream of new business ideas, a planetary structure works well. In

Figure 9.4: Application Teams Drawn From Corporate Functions

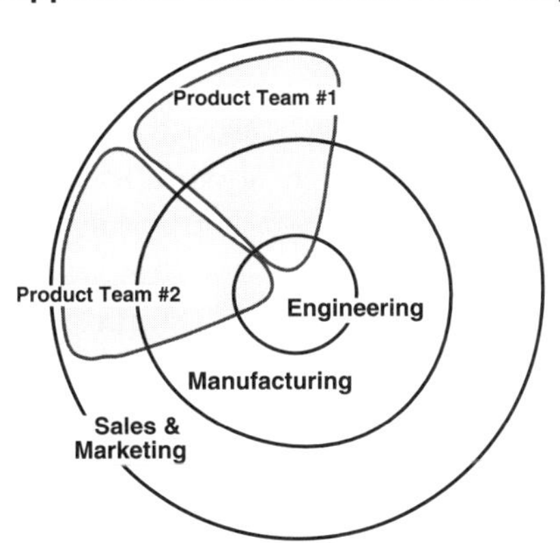

Figure 9.5: Planetary Systems

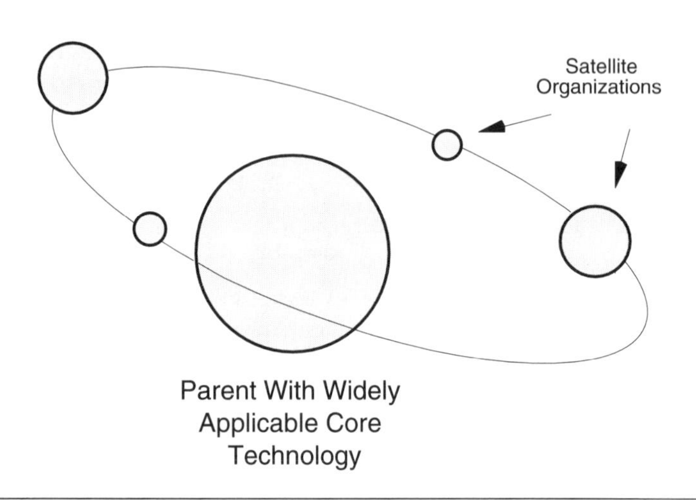

practice, the core functions almost as a think tank, conducting significant valued-added exploration, research, and development. This repository of the core technology may constitute an innovative technical center around which management can form smaller satellite organizations that operate independently but can regularly come back for new ideas (Figure 9.5). The satellites are separate business units of various sizes, developing and selling commercial applications, but they never stray far from the core. Each can develop its own structure.

New Technologies Unit

Large companies with established operations often prefer to put their new technologies into a separate division. Aerospace companies have done this for years, with Lockheed's famous "skunk works"—which conducted spy plane and experimental work for the military—being among the earliest. Setting up a new-technologies unit is a good technique for companies that believe the "new stuff" they are trying is different from what they do now and want to put it all together.

The new-technologies unit can be physically located on its own campus or in a locale remote from the rest of the business (Figure 9.6). The idea, of course, is to get the innovative people away from the bureaucracy of the larger establishment. The company needs innovations and wants them managed, not stifled. The

Figure 9.6: New Technologies Unit

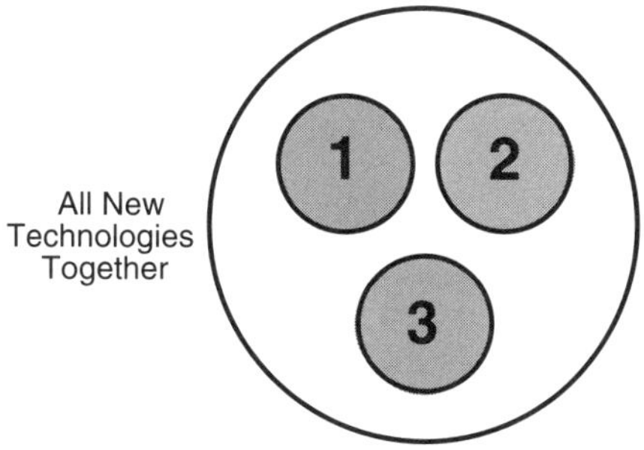

charter for such a unit can and should go beyond developing technology to include incubating new businesses.

A new-technologies unit is best headed by a mildly maverick vice president with a positive attitude and the managerial and financial knack for interacting with innovators. This manager is key to the success of a skunk works. He or she should have a wide comfort zone for people, disciplines, and developments and be capable of tracking numerous projects and efforts. Most important, this person needs the ability to keep technologists inspired and in touch with the realities of business, while keeping management optimistic and in touch with the realities of technology development.

Surround Yourself With Experience

As Chapter 6 pointed out, when a company with a technology or application needs access to a new market, an alliance or joint venture can provide the vehicle. (The corollary is also true: A company that has access to a market but needs a technology to license may well benefit from an alliance with a technology-rich outfit.) For the company with a valuable core technology and flexible production facilities, market-specific alliances can provide the arms and legs of distribution (Figure 9.7), leaving the business free to do what it does best: invent and produce.

Alliances help management wring the most value from applications at any stage. For instance, a company producing a specialty application can leverage its value through alliances with companies having access to markets with a need for that application, perhaps with some minor modification. If the company also

Figure 9.7: External Alliances and Joint Ventures

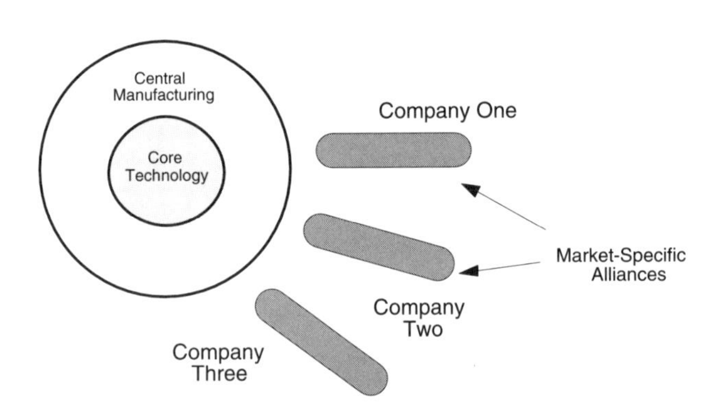

has a product moving to the commodity stage but retains a strong brand and access to the market, it can create a joint venture or license out the production and keep the marketing function. Instead of investing in productive capacity to chase fading margins, the company then rides the external producer's cost reduction curve while retaining a good portion of the value of the brand and market access.

These organizational models are neither exhaustive nor foolproof, but they enable you to provide environments that encourage innovation and nourish the seeds of change.

Technology Changes Everything

Technological innovation not only creates change, it also accelerates the tempo of the organization. If the outfit is sleepy, technology change wakes it up. If it is awake, technology gets it moving. If it is moving, technology moves it faster. Technology creates serendipity, starting change in new directions and creating unexpected new developments.

For a simple example of this, consider frequent flyer programs, a case where a touch of technology generated a tempest of change. American Airlines created the first frequent flyer program in 1979. The underlying technology amounts to computerized customer record keeping much like that of any electric utility or telephone company. The airline's computer simply records miles instead of kilowatts or minutes and generates premiums instead of bills. The technology had been there for years, waiting to be applied creatively.

Not only did American Airlines turn the liability of a record-keeping function into an asset, it turned it into an agent of change. The company dramatically changed its relationship with its customers. American gained the ability to win business loyalty in a regulated, commoditized industry on terms other than price and service. The new miles neutralized the effects of travel agents and corporate travel departments. The miles constituted a kind of currency that has since earned its place in our culture.

All of this arose from the ability to capture, store, track, and report the number of miles people flew on the company's planes.

In full-circle fashion, technology generates a need for new technology. It enables new scramble competitors to fashion substitutes for established products. But it also enables entrenched players to escape their ruts and move to new, higher-value applications. The extraordinary plasticity of technology, as manifested by bar codes, microprocessors, genetic engineering, and imaging, to name a few instances, guarantees that change will continue. It also guarantees that flexible, innovative, well-organized companies will always make money by applying technology to new applications and solutions.

Review and Preview

This chapter presents ways of dealing with the changes that technology and technology strategy invariably bring to a company. When you use technology for competitive advantage, you automatically enter the change business. The only question is how well you deal with it.

The first step is to anticipate change. The TAS and the adoption-diffusion model provide a glimpse of what lies ahead for a successful application. The tables in this chapter more specifically indicate what lies ahead for the company with a successful application. If you intend to take your outfit though the inevitable stages implied by each application zone, you face serious issues of organizational change. As the application moves to each new zone, you have to change many aspects of your company. The decision to do this must be made very carefully.

We have also looked at numerous tools for handling organizational change, including ten do's and don'ts for handling new-business initiatives. These tools included, above all, small work groups—TechnoLeverage Tip Number Nine is that *small work groups are the main tool for achieving technological change in business*—and seven organizational structures by which larger outfits avail themselves of the benefits of small groups.

With this chapter, we conclude our discussion of how to manage companies so as to achieve TechnoLeverage. In Chapter 10, we close by discussing how to invest in companies that have achieved it, pursuing both personal and professional benefits for the reader as an individual.

CHAPTER 10

The Winner's Circle

Investing in Leading-Edge Companies

In Chapters 1 through 9, we looked at how to use technology to lift a company's financial performance. Chapter 1 introduced TechnoLeverage and examples of companies that have used it. It also presented the technology applications spectrum, which charts the downward gross margin path of a successful application. Chapter 2 showed that some technology companies systematically achieve extraordinary returns, and it set forth their general business practices.

Chapter 3 examined the matter of searching for applications. Chapter 4 dealt with scramble competition. Chapter 5 discussed ways to take and hold the competitive lead in an application, a technology, or a market. We also looked at the benefits of leadership. Chapter 6 presented the tactics of TechnoLeverage: things that your company can do and steps you can take to fully exploit what you have and achieve TechnoLeverage.

Chapter 7 defined the contribution that financial managers can make to a business trying to leverage its technology and presented financial measures to manage and monitor such a company. Chapter 8 took you on a tour of technology, showing how to assess your company's technological resources and how to view the technologies available "out there" that might be of value to you. Finally, Chapter 9 examined the challenges of the changes that technology stimulates and how companies can best deal with those changes. That chapter stressed organizational structures that are useful to companies pursuing TechnoLeverage.

What we have learned in this book so far about technology strategy has practical application in two key areas: managing a company for high performance and investing in high-performance companies. The previous chapters showed how to use technology strategy to manage a company. This chapter shows how to select and invest in companies that use technology strategy. We cover a variety of investor perspectives, including those of the individual investor, the leveraged buyout participant, and the fund manager.

Recall that in Chapter 2 our research compared technology and nontechnology businesses and found slightly lower returns on assets for those in technology—a return on assets of 11 percent for technology companies versus 12 percent for nontechnology ones. Recall, too, that a small group of technology companies, 8 percent of the sample, were clustered around an 18 percent return on assets. These organizations tended to follow certain practices representing a model of how to manage a technology company. Their practices provided the seeds of technology strategy.

However, if past returns were a guarantee of future performance, investing in technology companies would boil down to finding those with an 18 percent return on assets, buying their shares, and promptly heading for the golf course. The challenge is to sort out the skilled TechnoLeverage companies from those merely lucky enough to be making a lot of money with a momentarily hot technology. There is also the matter of identifying nontechnology companies that achieve TechnoLeverage.

The investment question becomes: How do you find companies that use this particular model, those with the major elements we've presented in this book? As we've seen, these companies thrive in many industries, from the technology-intensive to the decidedly low-tech. As investors, we want to be able to recognize them, whoever and wherever they are.

Pair and Compare

I am a big believer in structured comparisons. When it comes to investment decisions, this means comparing pairs of companies in a similar industry, which we do in this section, as well as creating company peer groups, which we discuss later in this chapter. With pairs and peer groups, we are comparing qualitative characteristics and financial measures in the context of technology strategy. These comparisons help us identify companies having a technology strategy and the know-how and discipline to follow it. We seek investment opportunities in companies that have a technology and a model for managing it.

Our goal as investors is not to invest in "technology" or "a technology." Rather than invest in technology, our goal is the essence of TechnoLeverage Tip Number Ten:

> *TIP NUMBER TEN:* **Invest for the long run in technology plus management.**

We want to concentrate on technology companies that have learned how to reduce systematically the costs, risks, uncertainties, complexities, and high chance of failure that all technology companies face. We are looking for seasoned management teams that can minimize these factors—and for individual managers who know how to make money consistently with technology.

Comparing two businesses in the same industry illuminates the management dimension. Please understand that there is no quick, one-number litmus test to determine whether a company has a well-defined, long-term technology strategy. This chapter is not about shortcuts. It's about using good tools well. Investors must analyze each company carefully and compare it closely with others along a number of dimensions.

Beneath the Surface of Coatings

Two companies in the coated products industry enable us to make an in-depth, side-by-side comparison of the kind I recommend. Here is a pair whose qualitative and financial factors we can compare. Each of these examples is a division of a larger company. Let's call one Caliper Coatings and the other Riley Technologies. The individual profiles are disguised.

Both have superb technologies and world-class production operations. Moreover, as you will see, both have very healthy gross margins and sales per employee. Yes, superb technological prowess, world-class operations, high gross margins, and high sales per employee distinguish good businesses from mediocre ones. But these measures alone do not tell the whole story. We're looking for a pattern. Although both of these companies have superb technologies and attractive financial performance in several areas, only one has a business model that achieves TechnoLeverage. This model is key because as investors we are investing in the company's approach even more than in its technology.

Caliper Coatings develops proprietary applications in coated materials. It manufactures coated products for the paper and display markets, particularly ink-jet paper and paper for color ink-jet applications. Caliper combines a strong technological position, a strong brand, and strong relationships with well-known equipment manufacturers and important distribution channels. Its basic strategy is to produce leading-edge products for growth markets.

Riley Technologies also coats and converts papers and films, including slitting and sheeting, for the film processing industry. Riley and Caliper coat in similar ways. Essentially, Riley is a large, sophisticated job shop in that it provides a process—coating and converting services—to customers who have proprietary product technologies. Riley targets high-end industrial customers who value coating expertise or who need the capacity Riley provides, and it develops close partnerships with its customers.

In many ways these two are remarkably similar, so similar that if you did a cursory inspection you might have difficulty telling them apart for investment purposes. Table 10.1 compares the business models used by Caliper Coatings and Riley Technologies, the kinds of facts you'd normally glean from annual reports.

Although these companies have similar technologies and R&D budgets, they differ sharply in how they do business. Caliper provides proprietary *products* to a broad customer base in high-growth markets. Riley provides advanced (but not proprietary) *processes* to a narrow customer base in slow-growth markets. Caliper has more technological leverage and control over its destiny and better defenses against backward integration—that is, against customers deciding to do the work themselves—because its brand is strong and its products are distinctive. Caliper also takes the lead in developing technological advances. Riley has less leverage, since it provides a more generic (though high-quality) process that applies mainly to its customers' products and product technologies.

Table 10.1: Qualitative Comparison of a Pair of Business Models

Characteristic	Caliper Coatings	Riley Technologies
Product or service orientation	Makes products (branded and OEM).	Provides service almost exclusively.
Proprietary technology	Focuses on product (creates value-added products).	Focuses on process (does customized manufacturing).
R&D investment, annual	$7.1 million (5% of sales)	$6.5 million (4.4% of sales)
Technological advances	Many (materials chemistry)	Some (process improvements)
Manufacturing	World-class (makes and distributes own products).	World-class (produces products owned by customers).
Key markets	Concentrated, high-growth (color electronic imaging)	Numerous, mixed-growth (electronics, industrial, medical, displays)
Customer concentration	Broad base, 1,300+ (top 50 produce 82%).	Very narrow base, 45+ (top 3 produce 87%).
Sales and marketing focus	OEMs, distributor network, branded products	Selling direct (targets a few high-end, high-tech customers).
Control of destiny	High (can leverage own technologies into new products and markets).	Low (must rely on customers' technologies and needs for service).
Defense against backward integration	High (proprietary technology, unique products)	Low (two major customers have coating capabilities).

Table 10.2 summarizes the financial results produced by these two approaches.

By some routine measures, Riley is comparable with Caliper; Riley even has higher international sales, often thought of as desirable when world currency relationships are favorable. But there are clues, little ones—Riley's operating margin is about two-thirds of Caliper's, as is its profit per employee—as well as big ones. Look at asset turns and market growth to see Riley's deep problems and Caliper's significant advantage. Riley's return on assets of 8 percent (versus Caliper's 26 percent) and asset turnover of 1.2 times (versus Caliper's 2.2 times) say that Riley is simply not using its assets wisely. Management, and by extension the shareholders, have invested in too large an asset base for Riley's revenue, or they have invested in the wrong asset mix, or both.

Riley's overinvestment is bad enough, but its flat growth indicates that with current practices the company cannot grow its way out of its asset-heavy situation. Nor will rising market tides lift

Table 10.2: Comparison of Financial Characteristics and Results

	Caliper Coatings ($millions)	Riley Technologies ($millions)
Revenue	$142	$147
Gross margin	55.3%	53.8%
Operating margin	12.4%	8.9%
Return on net assets	26.1%	7.8%
Sales per employee	$264,500	$184,000
Profit per employee	$32,000	$20,000
Number of employees	558	801
Assets	$65	$122
Asset turns	2.2	1.2
Previous year's revenue growth	14.5%	-0.1%
Projected 1997 growth	13%	9%
1996 international sales/projected 1997 international sales	$8/$19	Both: $44
Growth rate of key markets	Blended market growth rate: 36%	Blended market growth rate: -2%

it passively out of its low-return hole. Production costs are not the problem, so Riley won't get far by cutting head count in production. Manufacturing efficiency is not the problem when Riley has a gross margin of 53 percent and sales per employee of $184,000. Riley's operating margin, especially in light of its high gross margin, does indicate the presence of high SG&A costs. However, Riley's biggest problems are too many assets and an inability to grow out of the situation, at least the way the company now does business. If Riley adds people or expenses to find new business, its net margins will suffer, reducing its return on assets even more. Financially, Riley is stuck, without a way to grow.

This comparison shows that investors in technology-based companies must examine the full array of qualitative and financial measures to piece together a picture of what is going on. Back in Chapter 7, we said that managers must do the same. If you go only by gross margin and sales per employee, coupled with the high quality of their work, you might be tempted to invest in Riley.

Meanwhile, Caliper Coatings generates almost the same revenue as Riley with a little over half the assets. Caliper's strategy of serving growth markets gives it plentiful growth opportunities.

Caliper's use of OEMs and distributors gives it leverage through market coverage. Caliper adds value directly and has invested in building its brand and technology. Given the information we have, Caliper would clearly be the preferable company to purchase for most individual investors and for most other kinds of investors—but not all.

Investors and Their Perspectives

How an investor views a company largely depends on the kind of investor he or she is. I am not talking about postures such as aggressive versus conservative. Although these factors affect investment decisions, I am instead referring to the role of the investor in relation to the company. Here are six investor perspectives through which we view the matter of investing:

1. Individual investors
2. Owners and directors of companies
3. Executives considering a management buyout
4. Corporate acquisition planners
5. Fund managers
6. Corporate managers

Individual Investors

If you are an individual investor following our system, you want a high return over a long period of time. This argues against what many investors in technology seek, which is a high return over a short time. Technology draws such investors because a technology company, particularly one with new, visible, successful applications, seems to have such an attractive profile. Yet many technologies do not produce what an investor might dream of. If you as an investor have a short-term technological horizon, you are going to wind up with a short-term investment horizon. A short-term investment horizon implies that you make numerous sequential investment decisions during a relatively short period. This places you in the position of having to track and correctly time what is poorly known even to those who buy the products themselves: the business prospects for numerous arcane, risky, fast-changing technologies.

The TechnoLeverage model says valuable things to the individual investor. It says that it is possible to find—and that you're much better off when you do find—one good stock you can hold for five years rather than five stocks to hold in sequence for one good year apiece. TechnoLeverage says that if you have good analytical tools, you can find high-return, long-term technology-plus-management investments rather than trying "serial bets on technologies." It says that although the movement and heat generated by many technology stocks encourage a frequent-trader style of investing, you can choose a less nerve-wracking approach and significantly improve your rewards from technology.

Remember, the more technology companies you buy—at once or in sequence—the closer to the average of technology companies your returns will be, and so the greater the chance you'll underperform the universe of nontechnology stocks.

TechnoLeverage says that you can pick companies with the right management. When you do, "your" management then deals with those arcane, risky, fast-changing technologies, leaving you free to play golf, fish, watch videos, or collect bottle caps. If you pick a company with restless, innovative, technologically opportunistic managers, you need not be a restless, innovative, technologically opportunistic investor; they do it all for you. They scan for new markets, develop new applications, monitor the financial controls, defend their core technologies, and do all the rest of it because *they have a model that calls for doing these things systematically*. Find these companies and you lock in the high returns of outfits that resemble the "bump" companies in Chapter 2. Find those managements and you avoid the one-trick ponies and technology rollercoasters that cause such indigestion.

This investment strategy is not about riding a rocket that could either soar or fizzle. Our strategy centers on finding a company that knows how to generate value and ignoring the ones that just might get lucky.

The oldest investment question in the world arises once again: How do you find those companies?

1. *Screen companies quantitatively on financials.* Look at gross margins, operating margins, sales and profit per employee, asset turnover, and return on assets, as outlined in Chapter 7. Look for high return on net worth with little or no debt, that superb indicator of the ability to self-finance. Collectively, these measures tell you whether the company is generating value.

2. *Screen for growth.* Look at growth in sales and earnings. Try to get a fix on the growth of the companies' largest markets and those of their customers. Judge growth projections by comparing them with past growth and with projections of peer group members. When looking at projections, remember that most businesses either are or are not innovative, marketing-oriented, and growth-oriented. Consider projections in light of the TAS. If a previously moribund outfit suddenly projects rapid growth, what has changed? Be skeptical of prophecies of sudden salvation.

3. *Screen for qualitative measures.* Once you have a pool of growing, value-generating companies, examine the source of the growth and value for each company. Ideally this should be a proprietary, applications-rich, hard-to-substitute core technology. Look for some basis for proprietariness. Look for patents, barriers to competition, or a variety of successful new products and a healthy number of markets served. Screen for breadth, depth, and size of markets. Look at their pricing strategy and the basis on which they compete.

4. *Examine their story.* Again, the reason I read annual reports is to determine which companies clearly define—with a story and with performance numbers—where the company has been, where it is going and why, and how it will get there. Those annual reports that do this tell you why to invest. Those that don't should steer you elsewhere. The company's story also emerges in the press and in management announcements regarding products, joint ventures, alliances, financings, and personnel and organizational changes.

The company's story is important beyond providing context and a rationale for the other items you are evaluating. It's important because the investment strategy I'm suggesting involves picking good stocks and holding them *until the story changes*. This differs from holding them until their stock-price performance deteriorates. (Then it's usually too late.) It means that a decision to sell should be driven by something more than an external event, even one that affects the price of the stock, because a long-term investment strategy in sophisticated technology management is by definition not event-driven. If management has a good technology and a good business model, they should be able to continue generating value, returns, and growth until something fundamental changes.

If the company's fundamental story and position does change for the worse, that's your sell signal. If senior managers leave the company or radically alter their strategy or just plain run out of high-value things to do, it's time to sell. Usually, you see these things coming over a period of months or quarters. The leadership thins out. The key numbers soften. Major customers and markets diverge. The standard gets superseded. The company starts venturing outside its bailiwick. It loses key scientists and can't replace them. When these things happen—and they often do—it is time to cash out.

Until then, there is no reason to sell unless you simply want the money for another purpose. If you have a good company, a good management, and a good stock, why cash out? So that you can go to the trouble of finding an unfamiliar new, equally good investment?

When it comes to investing in technology, the most common mistake individual investors make is to believe that they can somehow follow and judge the twists, turns, and future of a technology. That's not the investor's job; it's tough enough for those whose job it is. The goal of seeking TechnoLeverage businesses is to find managers who have figured out how to make money with technology so that you don't have to do the work yourself.

Owners and Directors

If you are the owner or a director of a company like Caliper, you are obviously in good shape. You have a good management team, and you should want to keep them. You also want to feed the company cash whenever necessary because you're getting a great return and the company is going to grow. Caliper is a durable profit-and-value-production machine.

If you are the owner or director of a company like Riley, you should forcefully press Riley to start using the tactics of TechnoLeverage. Riley's management lacks a high-return, growth-oriented business model. Yet they are not necessarily incompetent: Riley's gross margins are strong, its quality is world-class, and it's investing in R&D. It just isn't focused on growth and returns. It must be made to understand in the strongest possible terms that its future depends on the success of an all-out, no-holds-barred search for other, higher-value things to do—new applications, new markets, whatever it takes to get its assets working twice as hard.

A sell-off of some assets should be considered, but typically that's more easily said than done. Assets are rarely easily divisible. If there is some obvious perforation between Riley's low-return and high-return assets, wonderful. Sell off the low-return assets and you'll have your improvement. If such a perforation had existed, however, presumably management would have sold them off before asset turns dropped to 1.2 times a year.

Despite their positives, Riley's management could in fact be problematic. Riley's high gross margin coupled with the relatively low operating margin points to something wrong: With so few significant customers, where is the SG&A money going? Management's primary responsibilities are to earn a good long-term return and grow the business, and they now fail on both counts.

If you are the owner or director of both a Caliper Coatings *and* a Riley Technologies, you want to replicate, in Riley, the culture, values, and business practices of Caliper. An alternative would be to disinvest in Riley and invest more in Caliper. It's a matter of putting the shareholders' money where it yields the best returns and growth. The ultimate disinvestment in Riley would be to spin it off, perhaps to management (as discussed in the next section). At the parent-company level, a company like Riley can be a sinkhole for funds, time, and energy. It has almost twice the assets of Caliper, but no growth. To end the problem quickly, sell Riley.

In my experience, spin-offs are executed too infrequently and too late. Admittedly, there's an emotional cost to casting off a division. It is sad to part with people. It's also embarrassing to admit that you can't get a Riley to work and, perhaps, that it was a mistake to purchase it. Hope is another common emotion that surfaces repeatedly in these circumstances. Maybe Riley will get its act together. Sloughing off underperformers can take years. Meanwhile, the overall corporation is cash-constrained and stuck with underproducing assets. The sentimental approach underfeeds the more entrepreneurial divisions, starving the winners so that the losers can live.

If you own Riley, your best "partner" in a spin-off may be the organization's current management.

Executives Considering a Management Buyout

Riley Technologies could be an ideal candidate for a leveraged buyout (LBO) by management. There is a real possibility that the tactics of TechnoLeverage could work here. If management has the chance and they see a way to make these tactics work, they should buy out the owners. The assets are probably not worth any premium in a quick sale, given the turns and returns and the fact that their division hasn't been sold. If these assets are sufficiently bankable to the point that they can secure the debt needed to fund the buyout, an LBO would work nicely because the company is asset-heavy.

Management still needs ways of generating better returns and resuming growth, but there are many things they can build on if they have the motivation. They have to eliminate whatever problem is inflating the SG&A number. If their own salaries are the problem, they immediately have to take a reduction as well as defer income (the lenders will see to that). The managers must develop a desperate need for new customers and new markets—precisely the right signal! They need proprietary products, not just processes, and must pursue them fanatically. With all its faults, Riley represents an opportunity to add value to something that can probably be bought cheaply relative to its future value. In real estate terms, it's a fixer-upper. It has possibilities, and there is a plan to follow: Start with the core technologies, work outward, find problems to solve, develop applications and products, and take the new solutions to market—quickly.

On the other hand, again using real estate terms, Caliper Coatings may be the best house in the neighborhood. It would be expensive to buy. Corporate owners would be reluctant to sell unless the company didn't fit their game plan for some reason. In that event, they could still sell it outside, to a strategic buyer rather than to management, to get top dollar. It would be nearly impossible to buy as an LBO. Caliper does not have a large base of undervalued assets to secure the debt to finance an LBO. If you were looking for an LBO, Riley is actually the better bet.

Corporate Acquisition Planners

For a company planning an acquisition, either Caliper or Riley could make sense. Which one, if either, you would buy depends upon the motivations for the acquisition and the strategic fit between the acquiring company and its target.

Caliper Coatings is clearly a gem. But, as noted, gems are expensive. The price of Caliper, or an outfit like it, is strongly affected by whether it is a public company, privately held, or a division of a larger company. If a company like Caliper were a separate public company, it would be mighty expensive. The market bids the price of these companies up to high multiples of earnings and revenues. When you buy at a high multiple, you are betting strongly on future growth. However, if you see a clear path to growth, a strong proprietary technology, and a way to retain and motivate current management, even an expensive acquisition can make sense.

The cheapest way to buy a company such as Caliper is to purchase it from a multidivisional company or from private owners. You pay less buying from a corporation or private owners because these sellers seldom demand the inflated market multiple of a "growth stock." Rarely in a private purchase does the buyer pay a hot-growth-stock multiple of earnings, in the forties, fifties, or more; the exceptions are strategic buyers.

Nonetheless, even at a good price Caliper is simply too big for many potential purchasers. A Fortune 1000 company could easily handle it, but many smaller ones, not to mention individuals and partnerships, could not. If you are a small company or individual buyer, try to find an outfit as good as Caliper, with a similar financial and qualitative profile and similar returns and growth, but with one-fifth to one-tenth the revenue. A business with $20 million in annual revenue is much more within reach than a company with $200 million in revenue. Furthermore, if you choose correctly, you may get to enjoy some of the fastest revenue and earnings growth in the life of the company.

Riley Technologies is obviously not a gem. As discussed already, with the right management approach and some successful new applications and new markets it could be. But right now it is a fixer-upper, so concerns about overpaying are minimal. If it were a public company, Riley's stock price could well be below the book value of the assets. Since Riley is a corporate division, the parent company might be glad to be rid of it. So might a private owner. Therefore most of what would characterize a management LBO of Riley would apply to a corporate acquirer. The key necessity would be a sharply delineated, fact-based plan for rehabilitation and growth.

Fund Managers

Some fund managers might want to consider constructing a portfolio of companies that follow a TechnoLeverage model, especially fund managers who invest in technology companies. Some companies do consistently achieve the extraordinary returns that technology can generate. Although any particular technology can be unpredictable, these businesses have a methodical, stable approach to managing it. They have a model for repeatedly extracting and delivering maximum value from technology, and for generating high returns with it.

You can hardly do better over the long run than to "follow" business managers who fit this profile, repeatedly investing in the companies that they manage.

Even without building a true TechnoLeverage portfolio, technology fund managers should make room on their buy lists for companies whose strategies help them gravitate toward the 18 percent return on assets recorded by the high-performance TechnoLeverage companies. You may have to buy them at a high price, but over time they yield extraordinary returns compared with other technology companies and with the overall market.

The tactic of seeking companies in lower-tech industries that understand how to apply technology can also work for fund managers. Look at the long-term record of Nike or GE or FedEx, of Avery-Dennison or BankOne or MBNA or 3M. Such companies' management practices account for their success. They use technology as a strategic resource and know how to gain competitive advantage with it.

Fund managers can benefit from scanning to stay aware of the TAS, as well as from scrutinizing a company's technology strategy. Does the company have a valuable core technology or a valuable application for it? Does management have a method for creating value, or did they just get lucky? What dangers do encroachment and substitution (which we discussed in Chapter 4) pose to the company? When management comes to a decision point, what do they do? When the gross margin on a successful application declines, do they invest more to achieve production economies, or do they use leverage tactics? If they invest more when moving rightward, are they successful in the lower-margin application zones? If they use leverage tactics, are they successful in reaching a high-margin zone?

Finally, with regard to Caliper and Riley (if they were free-standing public companies), the fund manager, like the individual investor, should buy the former and avoid the latter, based on return and growth.

Corporate Managers

In a sense, of the six investor categories we're considering, the managers at a Caliper or a Riley are those most heavily invested in their respective companies. They invest their time, energy, and talent in them and no doubt have incentive compensation, stock purchase and 401(k) plans, retirement plans, and perhaps parent company stock options tied to performance. Thus, managers at both companies must do all they can to understand and apply TechnoLeverage.

Middle managers and some senior managers may wonder what they can do personally to raise their company's return or increase its sales. Traditional corporate strategies often provide little inspiration to middle managers and few practical ways for them to generate value and growth. A manager can't raise ROA by selling off the department's equipment and is not going to try increasing sales per employee by selectively firing portions of the staff without authorization. However, a real technology strategy enables almost all managers, regardless of their function, to overcome these limitations.

Any improvement counts. Everyone can scan for new problems, technologies, applications, markets, and customers. Each of these creates wealth for the company. Every individual in every function—executive, finance, accounting, marketing, sales, operations, engineering, R&D, service, and administration—can participate.

Technology strategy is therefore not something to be forced on people from the top down. Rather, it is a way of guiding—of instilling a value-maximizing approach in every employee—with the goal of eliciting each individual's full participation in the company and its future.

Peering Into Peer Groups

An investor in technology companies benefits greatly from constructing corporate peer groups. Corporate peer groups can be built on the basis of TechnoLeverage and other criteria. For example, a peer group can employ a technological approach as its primary criterion, and then include or exclude businesses based on other criteria such as size, growth rates, returns, and management strategy. Or it can use a criterion other than technology.

A peer group provides a broader basis of comparison than the side-by-side pairing we analyzed in the Caliper Coatings and Riley Technologies example. But these broader comparisons can be made along similar dimensions, such as performance, management effectiveness, and proprietariness of the technology. Also, a peer group based on a new technology can form an excellent basis for picking the Most Likely to Succeed, meaning the potential leader after a shakeout in that technology.

We have constructed peer groups for Caliper Coatings and Riley Technologies. Note that each company has its own peer group. We constructed these groups for a specific purpose and with selection criteria deliberately different from that of their shared technology. The actual purpose of these groups is to determine the value of Caliper and Riley independently, based on their circumstances and strategies, since their technical skills are so similar. The criteria we used for the Caliper Coatings peer group are:

- Revenue range: $60 million to $600 million
- Proprietary or patented product technology
- Potential for product branding
- High-growth markets

For the Riley Technologies peer group, we used:

- Revenue range: $60 million to $600 million
- Outsourcing of technical services
- Potential for using patented process technology
- Large, slow-growth markets

There is subjectivity in choosing peer-group criteria, and certainly in applying criteria such as "potential for product branding." That's fine; after all, you put together a peer group for your own purposes. Therefore, you decide which companies to include and on what criteria. Our goal is to construct a peer group based mainly on the business model followed by each company, which is in itself a subjective judgment. These criteria, applied to the universe of publicly traded companies in both the regular and expanded Value Line publications, produced these peer groups with these financial characteristics for the two companies (Tables 10.3 and 10.4).

Since our purpose is to set a peer-based valuation on Caliper and Riley, we are interested in developing a market capitalization factor to apply to the two companies. That factor is market capitalization as a multiple of sales (a measure discussed in Chapter 7). We developed this value for each of the six publicly held companies in the two peer groups. Then we calculated a size-weighted average of that figure for each group (with size based on sales). We applied the factor calculated for Caliper's peer group to Caliper and the one calculated for Riley's group to Riley. The result was the calculated market capitalization and the calculated market value per asset dollar for each company (Table 10.5).

Here's an investment difference! Table 10.5 says that for every dollar of assets invested in Caliper, its management creates $4.58 in market value. Table 10.5 also says that for every dollar of assets invested in Riley, 84 cents in market value is created. In other words, Caliper turns a dollar of assets into $4.58 worth of value, while Riley manages to destroy 16 cents of value on every dollar of assets in its hands. This method computes the value of Caliper

Table 10.3: Caliper Coatings: Business Model Peer Group

Company (ranked by revenue)	Description	Revenue 1995 ($millions)	Revenue Growth 1994-1995 (percent)	Operating Margin	Market Capitalization Early 1997 ($millions)	Market Capitalization as a Multiple of Sales
OM Group	Speciality chemicals for batteries, etc.	$361	43.8%	16.0%	$503	1.39
Acuson	Medical diagnostic systems	355	8.0	.5	644	1.81
Burr Brown	Analog integrated circuits for amplifiers	269	38.0	19.8	358	1.33
Lydall	Products with engineered fiber materials	252 (1996)	18.0 (1995-1996)	18.0	374	1.48
Safeskin	Hypoallergenic latex gloves	117	40.0	18.0	625	5.30
Ballard Medical	Patented medical products	81	24.0	37.5	504	6.10
Totals						
Revenue		$1,435				
Market capitalization					$3,008	
Market capitalization to revenue						2.10

Table 10.4: Riley Technologies: Business Model Peer Group

Company (ranked by revenue)	Description	Revenue 1995 ($millions)	Revenue Growth 1994-1995 (percent)	Operating Margin (percent)	Market Capitalization Early 1997 ($millions)	Market Capitalization as a Multiple of Sales
Jabil Circuit	Manufacturing outsourcing for circuit boards	$559	40%	5.1%	$308	.55
Dill Group	Electronics products and services outsourcing	335	60	11.0	184	.55
Layne	Contract oil field services	167	1	16.0	132	.79
IEC Electronics	Contract manufacturing	127	-2	21.8	45	.37
Benchmark Electronics	Contract manufacturing	97	-1	11.5	124	1.28
PCI Services	Integrated packaging services	86	7	19.5	168	.19
Totals						
Revenue		$1,371				
Market capitalization					$961	
Market capitalization to revenue						.70

Table 10.5: Valuation of Caliper Coatings and Riley Technologies, Based Upon Corporate Peer-Group Analysis

	Revenue ($million)	Peer Group Multiple of Sales	Calculated Market Capitalization ($millions)	Calcuated Market Value per Asset Dollar
Caliper	$142	2.10	$298	$4.58
Riley	147	.70	103	.84

Coatings at $298 million and that of Riley Industries at $103 million. Remember, these are two businesses with similar size, profits, technologies, and margins!

This approach to valuing technology-based companies reflects the fact that money is invested in a company to fund assets, that those assets are supposed to create value, and that the resulting value can be directly measured in dollars. In a public company, the stock market provides the valuation. For a privately held company or a corporate division, that value must somehow be calculated. The reason for using revenue here rather than earnings (as noted in Chapter 7) is that revenue is a relatively pure number: Revenue comes from real customers paying real dollars for what they see as real value. A lot can happen to revenue before it hits the bottom line; the top line directly reflects value as judged by customers buying the company's products—a good reason to employ revenue when calculating the value of a technology-based company.

Corporate peer-group comparisons represent a systematic, flexible, powerful way of approaching technology companies as investments. Regardless of your role as an investor, or of the ownership structure of the company, you can use this tool. Peer groups also work well when examining nontechnology companies. As with most tools, the longer and more frequently you use a peer group, the more powerful it becomes. Peer-group analysis forces you to look at a number of businesses, automatically challenging you to broaden your view and avoid falling in love with the first technology stock you see. Over time, you learn which companies manifest a strategy and which ones got lucky, which ones are buttoned up for the long haul and which are short-timers, which ones create value and which ones produce fireworks.

Invest in TechnoLeverage, Not in Technology

Think back to our list of cool technologies in Chapter 8. It is meant to raise your awareness of the technologies out there. The list is not, however, provided as an investment guide.

At one time or another, someone somewhere has gone wild over every one of the technologies on the list. Someone has pronounced each one of them to be The Future. Someone has looked at that technology and seen mountains of money. Fine, well, and good. But trying to buy one of these technologies is *not* the way to make an investment decision.

It's quite possible, even probable, that some investors will make millions in videoconferencing, low-orbit satellites, gene splicing, or super-pure products. Other investors will lose millions in the same technologies. Since you cannot forecast which outcome will occur, nor determine which company has the better technological approach, all you have is a glorified TechnoLottery.

As an investor you can't "buy a technology" and hope to have a winning strategy. To buy a technology in the true sense of the term, you would have to buy a substantial amount of stock in each company that participates in the technology. This means buying stock in enough companies to include at least several that turn out to be losers. This is like betting on every horse in a race at the track. It can't work. It can't work mathematically. And it can't work technologically, because you're betting your portfolio on something that could be superseded tomorrow. Somewhere a guy you've never heard of is working on some advance that comes from a direction you've never thought about.

What you can do is invest with business managers who have a proven model for making money with technology. Rather than try to find the next million-dollar technology to invest in, invest in these managers, in their companies, and in that model.

Review and Preview

In this final chapter, we have examined TechnoLeverage mostly from the outside looking in, that is, from the perspective of the investor rather than that of the manager. The goal of seeking TechnoLeverage companies is not to invest in technology. Instead, it is to find those select companies that, as stated by TechnoLeverage Tip Number Ten, allow you to *invest for the long run in technology plus management.* The investment goal is to find management teams that have a systematic model for creating value with technology and for capturing that value.

This implies a longer investment horizon than many people might associate with investing in technology. The recommended method aims to discover stocks that can be bought and held for an above-average long-term return. If you have good management with a good technology that it knows how to manage, there is no reason to sell unless you want the money or unless the company's story changes. If that happens, meaning if something fundamental has changed, then it is time to sell the stock.

This chapter has also discussed how to compare pairs of companies and how to construct peer groups for comparison purposes. Both of these tools provide context for evaluating companies from strategic, financial, management, and investment perspectives.

This book has laid out in detail a systematic way of looking at technology. It has portrayed technology as an accessible business resource that any creative and hard-working management team can press into service for greater growth and profit. Since technology drives so many business opportunities today, all businesspeople need a standard for evaluating technology so that they can accurately judge its business value. For any company that wishes to follow it, TechnoLeverage is the path to create and capture value, that is, to make money with technology.

APPENDIX

Tools for Knowing TechnoLeverage When You See It

Here are some guidelines to help you recognize TechnoLeverage when you see it in a company you manage, invest in, work for, buy from, sell to, or compete against.

Technological Guidelines

- Strong, application-rich core technology
- Proprietary technologies
- Patents
- Regulatory or other barriers
- Technologies applicable to own products
- Record of generating multiple applications
- Record of serving multiple markets
- Defenses against backward integration
- Repeated investment in technology
- Rapid development of technological solutions

Operational Measures

- Position of leadership
- Record of broadening the base of business
- Evidence of scanning and search strategies
- Preferred employer of technology experts
- Control of costs and wise use of capital

Operational Measures (continued)

- Incentive compensation for management
- Stable but flexible organizational structure
- Multiple strategic partnerships
- Multiple sales-and-distribution channels
- Reliable, well-priced suppliers
- Growth rate of major markets
- Proper positioning on technology applications spectrum

Financial Measures

- Gross margin
- Trend in gross margin
- Operating margin
- Net margin
- Return on assets
- Asset turnover
- Sales per employee
- Profit per employee
- Return on net worth
- Debt, expressed in weeks of sales
- Market growth
- Record of growth in sales
- Record of growth in earnings
- Market capitalization to sales relative to peer group
- Percentage of sales and profits from new products

INDEX

About Technology Marketing Group Inc.

Technology Marketing Group Inc. (TMG) is a management consulting firm that specializes in helping businesses manage and market technology. It is located in Acton, Massachusetts, and was founded in 1984. TMG works with senior managers to help them expand the value-added that their companies provide as well as help them find, evaluate, and enter new markets. Its website may be found at www.technology-marketing.com.

If you have thoughts, suggestions, or questions about the ideas in this book, you may send them directly to the author at mikehruby@technoleverage.com.